MELANIN

Edited by **Miroslav Blumenberg**

Melanin
http://dx.doi.org/10.5772/63257
Edited by Miroslav Blumenberg

Contributors

Magdalena Zdybel, Barbara Pilawa, Ewa Chodurek, Cecile Philippe, Markus Mitterhauser, Sandra Regina Pombeiro-Sponchiado, Gabriela Santana Sousa, Rita Cassia Ribeiro Gonçalves, Jazmina Carolina Reyes Andrade, Helen Cristina Favero Lisboa, Liviu Feller, Razia Khammissa, Miroslav Blumenberg

Published by InTech
Janeza Trdine 9, 51000 Rijeka, Croatia

Notice

Publishing Process Manager Iva Lipovic
Technical Editor SPi Global
Cover InTech Design team

First published February, 2017
Printed in Croatia
Legal deposit, Croatia: National and University Library in Zagreb

Additional hard copies can be obtained from orders@intechopen.com

Melanin, Edited by Miroslav Blumenberg
p. cm.
Print ISBN 978-953-51-2979-0
Online ISBN 978-953-51-2980-6

For EU product safety concerns:
IN TECH d.o.o., Prolaz Marije Krucifikse Kozulić 3, 51000 Rijeka, Croatia,
info@intechopen.com or visit our website at intechopen.com.

Meet the editor

Miroslav Blumenberg, PhD, was born in Subotica and received his BSc in Belgrade, Yugoslavia. He completed his PhD in organic chemistry at MIT; he followed this up with two postdoctoral study periods at Stanford University. Since 1983, he has been a faculty member of the R.O. Perelman Department of Dermatology, NYU School of Medicine, where he is a co-director of a training grant in cutaneous biology. Dr. Blumenberg's research is focused on the epidermis, expression of keratin genes, transcription profiling, keratinocyte differentiation, inflammatory diseases and cancers, and, most recently, the effects of the microbiome on skin. He has published over 100 peer-reviewed research articles and graduated numerous PhD and post-doctoral students. Dr. Blumenberg lives in New York, USA, with his wife and two children.

Contents

Preface

The word melanin refers to dark natural pigments produced by the oxidative degradation of tyrosine, catalyzed by tyrosinase, and polymerized into insoluble granular substance. The main function of melanin is to protect from harmful agents, primarily UV radiation, but also from oxidation, heavy metals, etc. In this volume, chapters deal with production of melanin in human oral mucosa (Liviu et al.), the regulation of melanin action (Cecile et al.), production and potential technological application of fungal melanins (Pombiero-Sponchiado et al.) and an innovative method for measuring melanin in various samples (Zdybel et al.)

In conclusion, this volume presents various biological and industrial aspects of melanin production, uses and analysis.

Miroslav Blumenberg
The R.O.Perelman Department of
Dermatology and Department of Biochemistry and Molecular Pharmacology,
NYU Langone Medical Center,
New York, USA

Section 1

Introduction

Chapter 1

Introductory Chapter: Melanin, a Versatile Guardian

Miroslav Blumenberg

Additional information is available at the end of the chapter

http://dx.doi.org/10.5772/67599

The word melanin, from the Greek word 'melanos' meaning black, refers to natural pigments produced by the oxidative degradation of the amino acid tyrosine into a variety of products that can self-polymerize into dark insoluble granular material. The first two steps of melanin biosynthesis are catalyzed by tyrosinase, a bifunctional enzyme that converts tyrosine first to dihydroxyphenylalanine (DOPA) and then to dopaquinone. The main function of melanin is to protect the organism from harmful agents such as ultraviolet (UV) radiation; melanin is capable of dissipating over 99% of absorbed UV light. Additionally, melanin can act as an efficient antioxidant, thus alleviating certain types of intracellular chemical damage, as well as protecting from the effects of high temperatures, heavy metals and oxidizing agents, and can play a role in the virulence of fungal pathogens by protecting them from the host's immune defense mechanisms [1, 2]. Certain organisms repurposed melanin for additional protective functions: ink of cephalopods, octopus, squid and cuttlefish, blackened by melanin, which provides protection from predators [3]. In mollusks other than cephalopods, melanin, together with other chromophores, serves to produce color patterns in their shells [4]. In insects, melanin is used even more resourcefully, not only for pigmentation of the exoskeleton but also for cuticle hardening, wound healing and in their innate immune responses [5]. Melanin in birds is responsible for coloration of the plumage and the exposed bare parts of the integument [6].

In humans, melanin is produced in specialized cells, the melanocytes, found in the basal layer of the epidermis. There are two basic types of melanin in human skin, eumelanin, brown-to-black, and lighter pheomelanin, found in people with red hair. Melanogenesis is enhanced after exposure to UV light, causing the skin to tan. Individuals with darker skin, having more melanin, are partially protected from UV-caused skin cancers. The distinct role of melanin in protecting our skin from deleterious effects of UV radiation has been recently reviewed [7]. Disorders of melanin production, whether acquired, such as vitiligo [8], or congenital, such as oculocutaneous albinism [9], are well-recognized human diseases.

Recently, importance of melanin, both in human diseases and in technological and industrial applications, attracted significant research interests [10]. Melanins, or rather various polydopamines, have found their use in materials science as a coating material for electronics and bioelectronics, in drug delivery systems, as sunscreens, as biofilms, etc. Melanin, thus, 'evolved' into a technologically important guardian as well. This necessitates development of a consensus among researchers on universally accepted and shared nomenclature, methodologies, standards and protocols in melanin research [11]. The current volume is a compilation of unpublished research results in this field.

Author details

Miroslav Blumenberg

Address all correspondence to: miroslav.blumenberg@nyumc.org

The Ronald O. Perelman Department of Dermatology and Department of Biochemistry and Molecular Pharmacology, NYU Medical School, New York, USA

References

[1] Karkowska-Kuleta J, Kozik A. Cell wall proteome of pathogenic fungi. Acta Biochim Pol. 2015;**62**(3):339–51.

[2] Nosanchuk JD, Stark RE, Casadevall A. Fungal melanin: What do we know about structure? Front Microbiol. 2015;**6**:1463.

[3] Derby CD. Cephalopod ink: Production, chemistry, functions and applications. Mar Drugs. 2014;**12**:2700–30.

[4] Williams ST. Molluscan shell colour. Biol Rev Camb Philos Soc. 2016;**23**(10):12268.

[5] Sugumaran M, Barek H. Critical analysis of the melanogenic pathway in insects and higher animals. Int J Mol Sci. 2016;**17**(pii):E1753.

[6] Galvan I, Solano F. Bird integumentary melanins: Biosynthesis, forms, function and evolution. Int J Mol Sci. 2016;**17**:520. doi:10.3390/ijms17040520.

[7] Natarajan VT, Ganju P, Ramkumar A, Grover R, Gokhale RS. Multifaceted pathways protect human skin from UV radiation. Nat Chem Biol. 2014;**10**:542–51.

[8] Manga P, Elbuluk N, Orlow SJ. Recent advances in understanding vitiligo. F1000Res. F1000 Faculty Rev-2234. 2016;**5**:1–9.

[9] Que SK, Weston G, Suchecki J, Ricketts J. Pigmentary disorders of the eyes and skin. Clin Dermatol. 2015;**33**:147–58.

[10] d'Ischia M, Wakamatsu K, Cicoira F, Di Mauro E, Garcia-Borron JC, Commo S, et al. Melanins and melanogenesis: From pigment cells to human health and technological applications. Pigment Cell Melanoma Res. 2015;**28**:520–44.

[11] d'Ischia M, Wakamatsu K, Napolitano A, Briganti S, Garcia-Borron JC, Kovacs D, et al. Melanins and melanogenesis: Methods, standards, protocols. Pigment Cell Melanoma Res. 2013;**26**:616–33.

Section 2

Melanin in Human Integumant

Chapter 2

Oral Mucosal Melanosis

Liviu Feller, Razia A.G. Khammissa and
Johan Lemmer

Additional information is available at the end of the chapter

http://dx.doi.org/10.5772/65567

Abstract

In the mouth, melanin is produced by melanocytes residing in the basal cell layer of the oral epithelium. Melanin influences the colour of the oral mucosa and provides protection against reactive oxygen species and bacterial-derived enzymes and toxins and acts as a physical barrier to both microorganisms invading the oral epithelium and to other microenvironmental stressors. The functional activity of epithelial melanocytes is regulated by biological agents in the microenvironment, including proopiomelanocortin-derived peptides, and by reciprocal interactions between melanocytes on the one hand and neighbouring keratinocytes and signals from the underlying lamina propria on the other hand. Oral mucosal melanin hyperpigmentation is common and may be physiological or pathological, and in either case the pattern of distribution and the intensity of the melanosis are variable. Physiological melanin hyperpigmentation is the result of increased melanin biosynthesis by melanocytes in the basal cell layer of the oral epithelium, but pathological melanin pigmentation may be the result of increased number of normal melanocytes or atypical melanocytes, of increased melanogenic activity of normal or atypical melanocytes, or of both. Oral mucosal melanin hyperpigmentation may be secondary to disease, medications, or smoking, and physiological oral melanin hyperpigmentation may be clinically and histopathologically similar so that the differentiation between pathological and physiological oral melanosis can at times be difficult.

Keywords: oral mucosa, melanocytes, melanosomes, melanotic macule, melanoma, naevus, physiological oral pigmentation

1. Introduction

There are both melanin-producing and amelanotic melanocytes in the oral stratified squamous epithelium, and the degree of melanin pigmentation of the oral mucosa is genetically determined. Dark-skinned persons more often have physiological oral mucosal melanin hyperpigmentation than do light-skinned persons. Regardless of race/ethnicity, the number and the distribution of melanocytes are very much the same within each oral anatomical region but vary from region to region. Even when there is no obvious pigmentation, oral mucosal melanocytes are always present and produce some melanin but the amount produced is determined by their melanogenic activity [1–4].

The vascularity, the blood haemoglobin concentration, the degree of keratinization of the epithelium, and the type and amount of melanin pigment present are variable determinants of the colour of the oral mucosa. The degree of melanin pigmentation *per se* is determined by the number and distribution of melanocytes in the basal cell layer of the oral epithelium of each anatomical region, by their melanogenic activity, by the degree of arborisation of the dendritic processes of the melanocytes, and by the number, size, and distribution of melanosomes in the 'melanocyte-keratinocyte unit' [4–6].

Most of the information on the physiology of oral mucosal melanocytes is derived from research on epidermal melanocytes. Oral mucosal and epidermal melanocytes are histologically and ultrastructurally similar, but under physiological conditions, it seems that the former are intrinsically less metabolically active. However, in response to chemical or physical environmental triggers, to certain drugs and hormones, to inflammation, and in association with certain systemic diseases or neoplastic processes, the metabolism of oral melanocytes may increase with consequent increase in biosynthesis of melanin [4, 6]. The intracellular and the microenviromental molecular signalling pathways that drive the development of oral mucosal melanin pigmentation remain largely unknown [7].

Brown/black eumelanin and yellow/red pheomelanin are pigmented polymers synthesized in unique organelles called melanosomes within melanocytes. Melanin has the capacity to scavenge reactive oxygen species and to neutralize microorganisms and their harmful products. Melanocytes also contribute to the homeostasis of skin and oral mucosa by acting as neuroendocrine cells secreting neuropeptides and as antigen presenting cells and phagocytes [4, 6, 8, 9].

Upregulation of melanin biosynthesis or increase in the number either of normal or of atypical oral mucosal melanocytes may bring about oral mucosal melanin hyperpigmentation. Physiological and pathological oral mucosal melanin hyperpigmentations may be similar in appearance, and sometimes may be difficult to differentiate even on the basis of history and microscopical features. Pathological oral mucosal melanin hyperpigmentations include those related to tobacco smoke, to certain systemic diseases or syndromes, to inflammatory processes, to certain drugs, and to developmental, benign, or malignant changes in melanocytes [4, 6, 10–13].

Engaging in any clinical or histopathological study on oral mucosal melanin hyperpigmentation is complex. Many subjects with oral mucosal melanin hyperpigmentation, whether physiological or pathological in origin, are usually unaware of it so cannot give information about its duration or progression. Oral mucosal melanin hyperpigmentation secondary to disease and physiological oral mucosal melanin hyperpigmentation may be clinically and histopathologically similar, and in some cases, it is almost impossible to determine whether oral mucosal melanin hyperpigmentation is physiological or is induced by smoking, by drugs, or by both.

Not all biological, physical, or chemical factors that may promote melanogenesis are known, neither are all the factors that may influence the size, distribution, or intensity of areas of oral pigmentation. Further research is needed to elucidate these issues; and into the roles that melanocytes and melanin play in the maintenance of homeostasis of the oral mucosa, as well as into the roles physiological or benign oral mucosal melanin hyperpigmentations may play as precursors of premalignant and malignant oral melanotic lesions.

In this chapter, we briefly review some aspects of the biology and physiology of oral melanocytes, the function of melanin in the oral epithelium, and the features of several different oral mucosal melanin hyperpigmentations, including physiological hyperpigmentation, tobacco-related melanosis, human immunodeficiency virus (HIV)-associated melanosis, melanotic maculae, naevomelanocytic naevus, and oral mucosal melanoma.

2. The biology of oral melanocytes

Melanocytes originate from neural crest precursor cells, which during embryogenesis migrate from the neural crest to their ultimate destination in the basal cell layer of the epidermis and of the epithelium of mucous membranes where they differentiate into mature melanin-producing melanocytes with complex network of dendritic processes [14]. However, some neural crest cells in their process of migration to the epithelium may become arrested in the lamina propria/dermis, remaining there in an immature state. These arrested immature melanocytes also sometimes termed naevomelanocytes, can become active, forming nests of cells giving rise to dermal/intra lamina propria naevi, and rarely to melanomata at those sites [15].

It is not clear whether the replacement of melanocytes of the basal cell layer of the oral epithelium lost to physiological processes of apoptosis or senescence, or to mechanical, thermal, or chemical injury is the result of cell division of 'mature' melanocytes in the basal cell layer of the oral epithelium which still retain their capacity of replication, or a result of migration from their local niche reservoirs of melanocyte stem/progenitor cells and their subsequent proliferation [6].

The mechanism by which the population of melanocytes of the oral epithelium is maintained in a steady state is unknown [7]. While melanocyte stem/progenitor cells and their reservoir niche have been identified in the skin, in the oral mucosa they have not been found [6, 15]. Tissue-specific melanocyte stem/progenitor cells are, however, relatively undifferentiated, divide infrequently, and have the capacity for self-renewal thus maintaining the genetic information necessary for regeneration of the population of melanin-producing melanocytes [16].

The ratio of melanocytes to keratinocytes in the basal cell layer of the oral epithelium, measured linearly, ranges from 1:10 to 1:15 [17]. However, this ratio varies at different oral mucosal sites of the same persons, but is similar in the same oral sites of different persons, regardless of their race/ethnicity [18, 19]. Melanocytes in the basal cell layer of the epithelium are connected to and communicate with their immediate neighbouring keratinocytes by means of tight junctions and gap junctions. Expression of E-cadherin cell adhesion molecules suppresses melanocyte proliferation, but on the other hand, expression of N-cadherin triggered by trauma or metabolic changes in the microenvironment promotes proliferation and migration of melanocytes, which can then aggregate in nests [4, 6].

It has been suggested that keratinocytes, through the release of biological mediators, proopiomelanocortin (POMC)-derived peptides, basic fibroblast growth factor, and endothelins, regulate proliferation and differentiation of melanocytes and stimulate melanogenesis through receptor-mediated signalling pathways [20].

Melanin is produced in melanosomes within the melanocytes in the basal cell layer of the oral epithelium. Mature melanin-containing melanosomes are transported from a perinuclear position in the cytoplasm of the melanocytes into and along microtubules towards the end of the dendritic processes, borne by the motor proteins kinesin and dynein. At the tip of the arborising dendritic processes, the melanosomes are transferred to neighbouring keratinocytes [4, 6, 21, 22]. It appears that each melanocyte is in contact with about 35 neighbouring keratinocytes in the basal and suprabasal cell layers of the epithelium, each forming a 'melanocyte-keratinocyte unit' [4, 6, 7].

3. Functions of melanocytes

Melanin influences the colour of the skin, oral mucosa, hair, and eyes. Melanin can also inactivate reactive oxygen species and free radicals, can sequestrate redox-active metals and organic toxic compounds, can neutralize bacteria-derived enzymes and toxins, and can downregulate inflammatory processes [4, 6, 7].

In the process of ascending through the cell layers of the epithelium to the surface desquamating layer, melanosomes lost from keratinocytes that form part of the melanocyte-keratinocyte unit disintegrate, releasing melanin 'dust' into the microenvironment. As melanin has strong binding properties, it can act as a physical barrier to both microorganisms invading the skin or mucous membrane and to other environmental stressors [4, 6, 7].

Although melanin itself provides protection against reactive oxygen species (ROS) and toxic free radicals, paradoxically, the biosynthesis of melanin itself generates ROS that may cause DNA damage; and quinones and semiquinones, which are intermediates of melanin biosynthesis, are mutagenic with the capacity to cause cytogenetic instability. Therefore, loss of the integrity of melanosome membranes with leakage of these intrinsic toxic agents has the potential to cause self-injury to the melanocytes [6].

Apart from the beneficial effects and detrimental side effects of melanin production, melanocytes can also protect against microbial infections by acting as antigen-presenting cells producing cytokines, phagocytosing microorganisms, and degrading both phagocytosed bacteria and bacteria in the microenvironment by means of melanosomal lysosomal enzymes [4, 6, 7].

Melanocytes can also function as neuroendocrine cells by producing acetylcholine, catecholamines, and POMC-derived peptides, which have the capacity to mediate local immunoinflammatory and antimicrobial responses and to modulate the biological behaviour of oral tissues [4]. Melanocytes function as stress sensors, as immunocytes, and as neuroendocrine cells, thus playing several important roles in maintaining tissue homeostasis [23–25].

The functional activity of epithelial melanocytes is regulated by the reciprocal interaction of melanocytes in the basal cell layer of the epithelium with the underlying lamina-proprial connective tissue and by several other biological agents. These include the POMC-derived peptides released by neighbouring keratinocytes and the neuropeptide calcitonin gene-related peptide, substance P, and neuropeptide Y released from intramucosal free nerve endings, which are in close contact with melanocytes in the basal cell layer of the epithelium. Adrenergic and cholinergic agents, growth factors, inflammatory mediators, and other autocrine and endocrine stimuli also contribute to the regulation of the functional activity of melanocytes [6].

4. Melanin biosynthesis

Premelanosomes contain all the proteins and enzymes required for their maturation to the fully functional specialised membrane-bound organelles called melanosomes, which are the sites of biosynthesis of melanin [26, 27]. The fully functional melanosomes contain all the proteins and enzymes required for biosynthesis of melanin, and the structural matrix proteins form an internal scaffold both to support the architecture of the melanosomes and to serve as a nidus for melanin deposition [22].

The melanosomes in melanocytes residing in the basal cell layer of the epithelium can synthetise two chemically distinct types of melanin: brown/black eumelanin and red/yellow pheomelanin, both of which are derivatives of the amino acid tyrosine through the intermediate step of conversion of dopa to dopamine [4]. Both eumelanin and pheomelanin can be produced in the same melanocyte, but by different melanosomes. Eumelanosomes and pheomelanosomes differ in shape and in their patterns of internal melanin deposition. Eumelanosomes are elliptical and characterized by longitudinal deposition of the pigment; and pheomelanosomes are spherical, characterized by granular deposition of the melanin [4]. The ratio of eumelanin to pheomelanin is genetically determined by a number of factors including the degree of base-line functional activity of the enzymes and proteins driving the 'mixed melanogenesis', by the melanocortin 1 receptor (MC1R) genetic polymorphism, and by the amount of the tyrosine in the melanosome [6].

The following agents are important in melanin biosynthesis. Tyrosinase drives the complex conversion of tyrosine to melanin. p protein is involved in stabilizing the melanosomal protein complex, in regulating the melanosomal pH, and in transporting the requisite proteins to the melanosomes. Tyrosinase-related protein 1 stabilizes the enzyme tyrosinase and is involved in melanosome maturation, and tyrosinase-related protein 2 modulates the quantity and quality of the melanin produced. The membrane-associated transport protein (MATP) functions as its name implies [5, 28].

5. Regulation of melanin biosynthesis and distribution

Upregulation of melanin biosynthesis may be triggered by external stimuli, such as radiation, tobacco smoke, certain drugs, and endogenous paracrine and endocrine agents. POMC-derived peptides including adrenocorticotropin (ACTH), alpha melanocyte stimulating hormone (α-MSH), β-MSH and β-endorphins, MC1R, adrenergic and cholinergic agents, and growth factors, cytokines, or nitric oxide (NO) secreted in the local microenvironment are all involved in the regulation of melanogenesis [6].

POMC-derived peptides, particularly α-MSH, stimulate MC1R of melanocytes, releasing the intracellular second messenger cAMP, which in turn induces a cascade of intracellular molecular events, ultimately activating the microphthalmia-associated transcription factor (MITF). This is the master regulator of genes driving the process of melanogenesis. Thus, the α-MSH/MC1R/cAMP/MITF molecular signalling pathway is fundamental to melanin biosynthesis and to a great extent determines the type and amount of the melanin produced [4, 6]. The MC1R gene is highly polymorphic; the most common MC1R variant promotes eumelanin production, but other germline variants are associated with increased pheomelanin and decreased eumelanin biosynthesis [6].

β-endorphin, the POMC-derived opioid peptide, has the capacity to upregulate melanogenesis and melanin distribution by promoting maturation of melanocytes and arborisation of their dendritic processes, thus increasing the efficacy of the transfer and distribution of melanosomes to neighbouring keratinocytes [23, 29].

Non-neural adrenergic and cholinergic signalling pathways are expressed by melanocytes. Activation of adrenalin/β2-adrenoreceptor/cAMP/MITF and of noradrenalin/α1-adrenoreceptor/cAMP/MITF also induces melanin biosynthesis and arborisation of melanocytic dendrites. Activation of nicotine receptors on oral melanocytes by acetylcholine or by nicotine may bring about melanin hyperpigmentation [4, 6, 30].

Inflammatory mediators can induce production of prostaglandin E_2 (PGE_2) and NO, which in turn have the capacity to promote maturation and increase transport of melanosomes, to upregulate the expression of the tyrosinase gene, and to increase the complexity of the network of melanocytic dendrites. PGE_2 and NO together promote melanogenesis and melanin deposition [4, 6, 31, 32].

To the best of our knowledge, the role that the melanocortin, adrenergic, and cholinergic systems play in relation to melanogenesis in the oral mucosa has not been demonstrated, and current knowledge of these systems is from research on epidermal melanin.

6. Oral mucosal melanin hyperpigmentation

Oral mucosal melanin hyperpigmentation is common and may be physiological (racial) or pathological [4, 6]. In either case, the pattern of distribution and the intensity of the pigmentation are variable. The pigmentation may be the result of an increase in the number of melanocytes in the basal cell layer of the oral epithelium, or of increased melanin biosynthesis by the melanocytes. Sometimes pathological hyperpigmentation may be brought about by hyperplasia or increased melanogenesis of atypical melanocytes [33]. Melanogenically active melanocytes/nevomelanocytes in the lamina propria also have the capacity to cause oral mucosal melanin hyperpigmentation [34]. Occasionally histopathological examination of a biopsy specimen from a pigmented lesion is needed to rule out a malignancy.

6.1. Physiological oral mucosal melanin hyperpigmentation

Physiological oral mucosal melanin hyperpigmentation manifests clinically as asymptomatic, single or multiple, well-demarcated or ill-defined patchy or uniform macules which range in colour from light to dark brown or black, and are of variable size and configuration. It may affect any part of the oral mucosa, but most frequently the gingiva, where it is usually bilaterally symmetrical, does not transgress the mucogingival junction, and does not involve the marginal gingiv (**Figure 1**) [4, 6]. Physiological gingival melanin hyperpigmentation is often more pronounced in the anterior than in the posterior part of the mouth, and the buccal/labial surfaces are more intensely pigmented than the lingual/palatal [35].

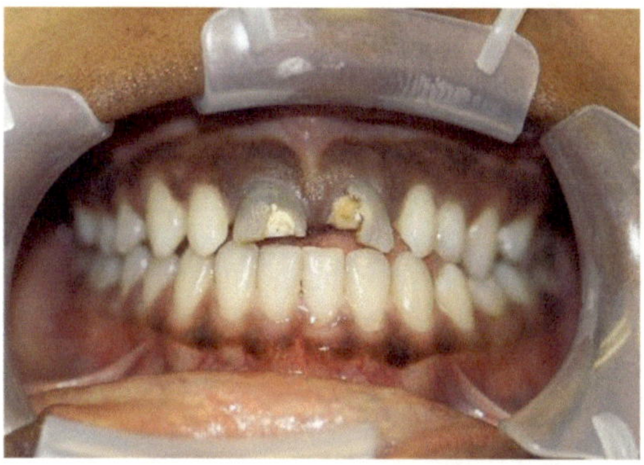

Figure 1. Generalized black-brown physiological melanin hyperpigmentation of the maxillary and mandibular gingiva, not transgressing the mucogingival junction. The patient's main concern was her carious incisors.

Physiological oral mucosal melanin hyperpigmentation is very common in Blacks, in general is more frequent in darker-skinned than in lighter-skinned persons regardless of their race/ethnicity, and is most probably caused by genetically determined metabolic hyperactivity of oral melanocytes. It affects males and females equally and usually develops during the first three decades of life. Many persons with oral mucosal melanin hyperpigmentation are unaware of its presence [7]. According to some publications, the extent and the intensity of physiological oral mucosal melanin hyperpigmentation increase with age probably owing to the cumulative effects of endogenous and extrinsic melanogenic stimuli, such as inflammatory processes, drugs, tobacco smoke, and recurrent minor functional injuries [4, 6, 11, 13].

In physiological oral mucosal melanin hyperpigmentation there is no increase in the number of melanocytes, but there is increased melanin in melanocytes in the basal cell layer of the oral epithelium, in the 'melanocyte-keratinocyte units', and in the lamina propria within melanophages. Occasionally, extracellular particles of pigment may be observed. These microscopical features are very similar to those observed in idiopathic melanotic macules, in melanosis related to smoking tobacco or inflammation, in melanosis induced by certain medications, and in HIV melanosis [11, 36]. Therefore, a thorough medical history may be helpful for differential diagnosis, but the differentiation of physiological oral mucosal melanin hyperpigmentation from single or multiple melanotic macules based on medical history and on clinical and histological grounds may be impossible as they share the same clinical and microscopical features and both may develop at a young age. It is also almost impossible to determine whether the oral mucosal melanin hyperpigmentation in an HIV-seropositive smoker who is on multiple drug therapy is owing to the medication, to the HIV infection, to the use of tobacco, or to the interaction between these factors.

Physiological melanin hyperpigmentation does not require treatment, but if for cosmetic reasons it has to be removed, laser or cryotherapy or surgical excision are often successful. However, recurrence is not uncommon most probably because treatment does not always completely eliminate the genetically programmed hyperactive melanocytes in the epithelial field surrounding the hyperpigmentation [4].

6.2. Inflammation-associated oral mucosal melanin hyperpigmentation

Under circumstances that are not well defined, certain inflammatory cytokines and mediators have the capacity to promote the production of α-MSH by oral melanocytes and keratinocytes and to upregulate the expression of MC1R of melanocytes resulting in increased biosynthesis of melanin. Independently of this α-MSH/MC1R pathway, PGE_2 can promote tyrosinase activity and arborisation of melanocytic dendrites. Thus, melanin hyperpigmentation may develop at chronically inflamed oral mucosal sites, as in association with repetitive local trauma or injury or with immunoinflammatory pathological conditions, such as oral lichen planus and oral lichenoid reactions [37].

Such inflammation-associated oral mucosal melanin hyperpigmentation occurs most commonly in darker skinned persons and usually manifests as single or multiple brown-black patches, which may persist for a long time after the resolution of the inflammatory process [6, 11, 37]. These inflammation-associated hyperpigmented patches are of no clinical significance, their biological significance is unknown, and treatment is not required.

6.3. Tobacco-related oral melanosis

Tobacco smoking may cause diffuse brown-black hyperpigmentation of the oral mucosa in 20–30% of chronic heavy smokers. This usually affects the gingiva, palate, and the buccal mucosa, and the intensity of the pigmentation is related to total tobacco usage. It has been suggested that the melanin produced in response to cigarette smoking is a protective reaction, contributing to the detoxification of polyaromatics, nicotine, and benzopyrene in the tobacco smoke. Smoker's melanosis may gradually diminish if the smoking habit is stopped. Importantly, smoker's melanosis does not undergo malignant transformation [11, 36].

As in differentiating some other cases of oral mucosal melanin hyperpigmentations one from another, on clinical and histological grounds alone it is impossible to make a positive diagnosis of tobacco-related melanosis.

6.4. Human immunodeficiency virus (HIV)-associated and medication-induced oral mucosal melanin hyperpigmentations

Oral melanin hyperpigmentation is common in HIV-seropositive subjects. It may develop secondarily to HIV-induced cytokine dysregulation, to medications used in the treatment of HIV infection or HIV-associated diseases (zidovudine, clofazimine and ketoconazole), or to adrenocortical deficiency, which is not infrequent in HIV-seropositive subjects with a low CD4+ T cell counts [6, 10].

It is unknown whether structural proteins of HIV can stimulate melanocyte activity directly to upregulate their melanin biosynthesis. It is possible that HIV-associated mucosal hyperpigmentation may fortuitously represent a local protective immune reaction against subclinical oral infections and concomitant inflammatory processes. Indeed, it has been suggested that certain cytokines that are upregulated during HIV infection have the capacity to induce the production of α-MSH by oral melanocytes and keratinocytes and to mediate the upregulation of expression of MC1R by melanocytes, resulting in increased melanin production. In some HIV-seropositive subjects, this may trigger oral mucosal melanin hyperpigmentation [6, 14].

If it occurs, HIV-associated oral mucosal melanin hyperpigmentation usually develops within 2 years of contracting HIV or within a few months of starting antiretroviral treatment with the drug zidovudine. Typically, it manifests clinically either as solitary or multiple dark melanotic macules (**Figure 2a, b**) or as diffuse brownish or brown-black areas of melanin hyperpigmentation (**Figure 3**). This may involve any part of the oral mucosa but most frequently the buccal mucosa, affecting females and males equally [6, 14].

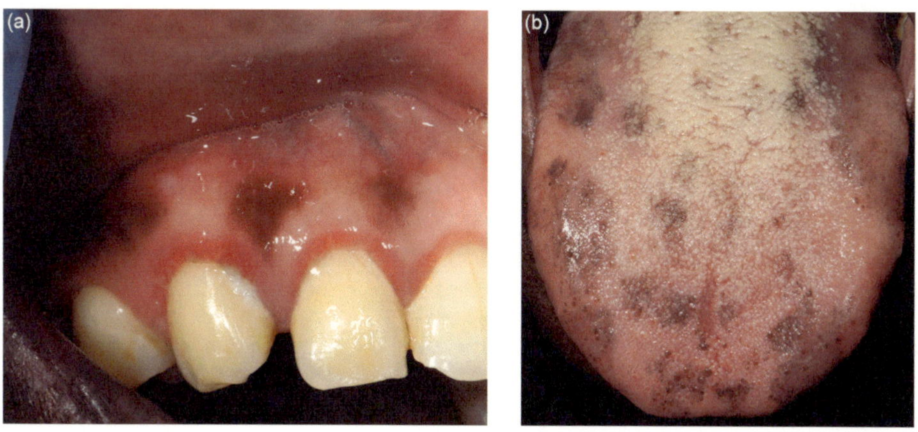

Figure 2. (a) Multiple pigmented maculae of the maxillary labial attached gingiva in a 32-year-old HIV-seropositive female with a CD4+ T cell count of 425 cells/mm^3. (b) Multiple pigmented maculae of the dorsum of the tongue in a 42-year-old HIV-seropositive female on HAART with a CD4+ T cell count of 176 cells/mm^3.

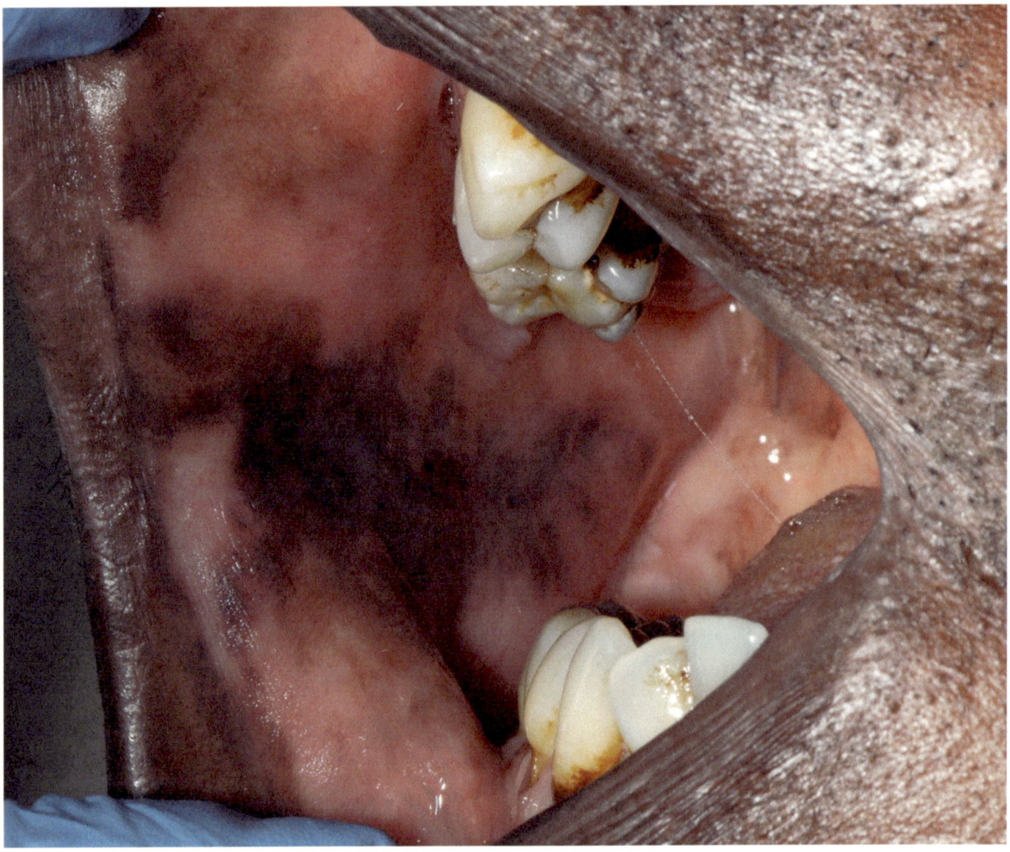

Figure 3. Irregular, diffuse, mottled hyperpigmentation of the buccal mucosa in a 40-year-old HIV-seropositive male with a CD4+ T cell count of 240 cells/mm^3.

Apart from zidovudine used in the treatment of HIV infection, antimalarials, oestrogen, ketoconazole, clofazimine, and imatinib may mediate the development of oral mucosal melanin hyperpigmentation [12, 38]. There will usually be a gradual diminution of the hyperpigmentation after the medication is discontinued [6].

The mechanisms by which certain medications bring about oral mucosal melanin hyperpigmentation include upregulation of tyrosinase activity, deposition of complexes of the drug with pre-existing melanin in the mucosa, or induction of local inflammatory reactions with the triggering of melanin production [11, 38].

Histopathologically, medication-induced melanin pigmentation is characterised by the presence of melanin within the lamina propria either as free granules or within melanophages, but without melanocytic hyperplasia. Melanosis of basal cells can commonly be observed [11, 38].

http://dx.doi.org/10.5772/65567

6.5. Oral mucosal melanin hyperpigmentation associated with syndromes and systemic diseases

Oral mucosal melanin hyperpigmentation is often observed in the systemic conditions such as Peutz-Jegher syndrome, McCune-Albright syndrome, Laugier-Hunziker syndrome, Addison disease, and neurofibromatosis. The clinical appearance is one of the brown to black spots or macules, with a histopathological increase in melanin in the basal cell layer of the oral epithelium and melanin incontinence in the upper portion of the lamina propria, but without an increase in the number of melanocytes [11, 39, 40].

6.6. Oral melanotic macules

An oral melanotic macule, by conventional definition is a focal, well-defined, uniformly coloured oral mucosal hyperpigmentation, less than 1 cm in diameter, of unknown aetiology. The colour of the macule may range from light to dark brown. Any part of the oral mucosa may be affected, but particularly the buccal mucosa, and although they are usually solitary, there are sometimes several oral melanotic macules. The average age at diagnosis is 43 years, and it is observed more commonly in females than in males [11, 13, 36, 40]. Any recently developed oral melanotic maculae that are irregularly pigmented or have recently increased in size should be viewed with suspicion and microscopically examined to exclude melanoma; and melanotic maculae of the maxillary gingiva or the palatal mucosa which are the most common sites of melanoma should be viewed with even greater suspicion [34, 41].

6.7. Oral melanoacanthoma

Oral melanoacanthoma, mainly seen on the buccal mucosa of young Black women, is an uncommon, asymptomatic, brown to black melanotic lesion, reactive in origin, usually flat or slightly elevated with a smooth surface. It usually increases rapidly in size to several centimetres and is characterised by acanthosis and spongiosis of the affected epithelium with dendritic melanocytes dispersed throughout the thickness of the epithelium, with a mild inflammatory cell infiltrate in the superficial lamina propria [11, 13, 34]. As the appearance of oral melanoacanthoma is not diagnostic, diagnosis must be established histopathologically.

6.8. Oral mucosal melanotic naevus (naevomelanocytic naevus)

Broadly, the term oral naevus refers to a congenital or acquired melanotic pigmentation of the oral mucosa brought about by abnormal excessive accumulation of melanocytes/naevomelanocytes at the junction of the epithelium and the lamina propria or in the lamina propria [34]. The sequence of biological events leading to the development of oral naevomelanocytic naevi is largely unknown. It is suggested that oral naevomelanocytes, like oral melanocytes, are derived from neural crest precursor cells, which during embryogenesis migrate to a final destination in the basal cell layer of the oral epithelium. However, it is not clear which one of several mechanisms are implicated in the development of melanocytic naevi. Naevomelanocytes may originate from melanocyte precursors that have acquired some genetic alterations during development, consequently preventing them from differentiation into mature functioning melanocytes in the basal cell layer of the oral epithelium; they may originate from mature melanocytes residing in the basal cell layer of the epithelium that have undergone some cytogenetic alterations culminating in their dedifferentiation into naevomelanocytes or they may originate from stem/progenitor melanocytes that have undergone cytogenetic alteration in the process of replacing mature melanocytes in the epithelium lost to mechanical, thermal, or chemical injury or as a result of apoptotic processes. A further possible mechanism is that naevomelanocytes originate from neural crest cells destined to become melanocytes in the oral epithelium, but for reasons unknown remain entrapped in the submucosa. Naevomelanocytic naevi may thus be manifestations of a developmental malformation or of hyperplasia of melanocytes [34, 42].

As many naevomelanocytes in cutaneous naevi show activating mutations in the BRAF or in the NRAS intracellular signalling pathways which mediate cell proliferation and differentia-tion [11, 34, 42], naevi are considered by many to be benign naevomelanocytic neoplasias. Histopathologically, naevomelanocytic naevi are characterised by several directly contacting naevomelanocytes in clusters confined to the junction of the epithelium and the lamina propria (junctional naevi), to both junctional zone and to the superficial lamina propria (compound naevi), or only to the subepithelial connective tissue (intra-lamina proprial naevi). The naevomelanocytes within all three types of naevi are devoid of melanosome-transferring dendritic processes, so that the colour of these naevomelanocytic naevi derives entirely from the melanosome content of the melanocytes themselves. Most oral mucosal melanocytic naevi are acquired rather than congenital and are of the intra-lamina proprial type [11, 13, 34, 36, 43].

Acquired oral mucosal melanocytic naevi are uncommon, occur more frequently in females than in males, and are observed mainly on the hard palate, gingiva, and buccal mucosa. The mean age at diagnosis is 35 years. They are small, well-defined asymptomatic macules or papules, usually brown to black in colour (**Figure 4**), but sometimes may be bluish-grey [11, 13, 36, 40].

The blue naevus is a less common oral mucosal naevus. The clinical appearance is of small blue macules or papules almost always on the palates of children or young adults (**Figure 5**) [34] and is characterised histopathologically by spindle-shaped melanin-producing naevomelanocytes arranged in fascicles parallel to the overlying epithelium, deep within the lamina propria [11, 13, 34].

6.9. Oral mucosal melanoma

Oral mucosal melanoma is an uncommon malignancy accounting for only about 0.5% of all oral cancers. It is rapidly growing and usually asymptomatic, so it is often diagnosed only when the melanomatous lesions are already advanced and there has been metastatic spread to the regional lymph nodes. Under these circumstances, the prognosis is poor with a mean 5-year survival rate of 15–20% [15, 44]. Large clinical size, microscopical evidence of deeply invasive front, a high mitotic rate of the melanoma cells, lymph node metastasis, and vascular or neural invasion and spread are all indicators of poor prognosis [45, 46].

http://dx.doi.org/10.5772/65567

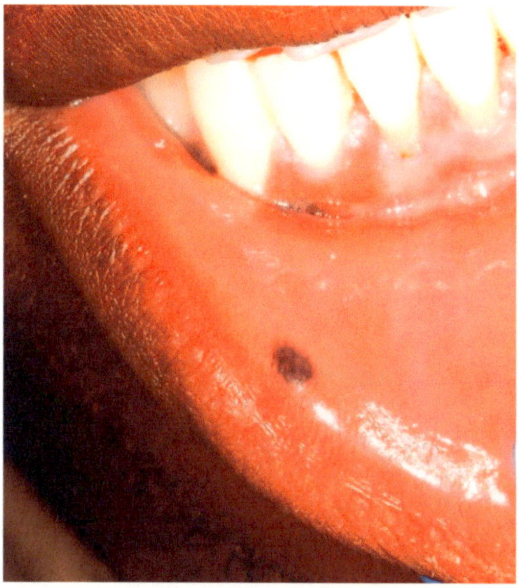

Figure 4. A 20-year-old male had this melanotic lesion of the lower labial mucosa, self-reportedly since birth. It was probably a congenital naevus but the patient refused biopsy. Note also the physiological gingival melanosis.

The palate and the gingiva are the oral mucosal sites most frequently affected. Up to one-third of oral mucosal melanomata arise from pre-existing benign oral mucosal hyperpigmentations [15, 47–49] and the remaining two-thirds arise *de-novo* [15, 44, 50, 51].

Oral mucosal melanoma differs from cutaneous melanoma in its profile of cytogenetic alterations, its more aggressive clinical course and the fact that it is not associated with any known carcinogenic agents or extrinsic risk factors [15]. The oral lesions are usually irregularly shaped melanotic macules, papules, plaques, or a combination of these forms, and their pigmentation is non-uniform, with mottled shades of grey, dark blue, dark brown, or black (**Figure 6a–c**) [15, 44].

The melanoma precursor most probably originates from a stem/progenitor cell that has undergone cytogenetic alterations and consequently expresses dysregulated developmental signalling pathways and transcription factors, ultimately acquiring a malignant phenotype. However, it is also possible that melanoma precursor cells originate from mature melanocytes in the basal cell layer of the epithelium that have undergone dedifferentiation following cytogenetic alterations or from immature melanocytes/naevomelanocytes arrested in the submucosa during migration from the neural crest which for reasons unknown have acquired a malignant phenotype [15].

The acquired genetic alterations confer upon the initially transformed atypical melanocytes and later upon the melanoma cells that evolved, a selective advantage over their neighbouring normal cells in terms of fitness and proliferation. Melanoma cells then have the capacity

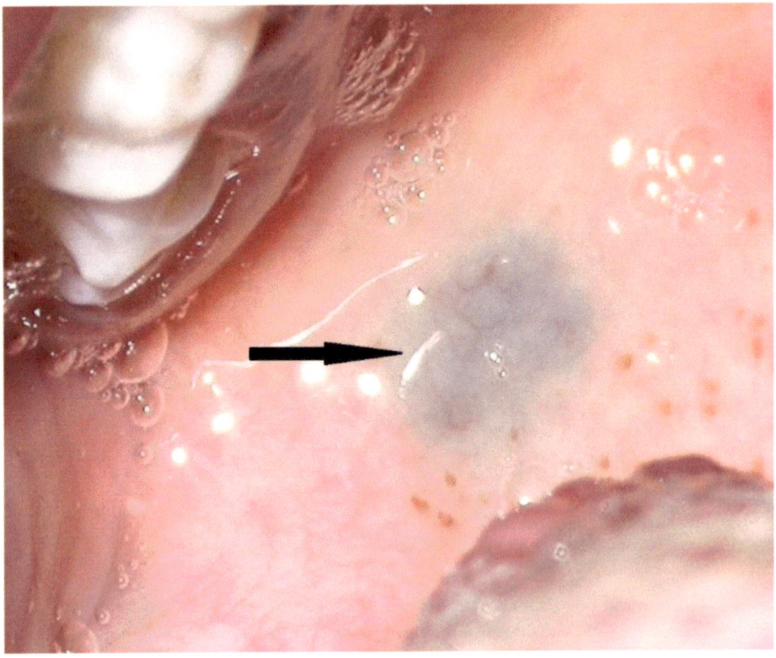

Figure 5. A blue naevus (arrow) on the soft palate in a 35-year-old male. Patient was unaware of its presence, and it was observed on routine oral examination. Histopathological examination confirmed the clinical diagnosis of blue naevus.

to undergo clonal expansion to invade and destroy local tissues, to metastasize to regional lymph nodes, and to spread to the lungs, bone, liver, brain, or skin [15, 52–54].

Melanoma cells arising from the immature melanocytes that have become arrested in the lamina propria of the oral mucosa proliferate, forming nodular aggregates in the lamina propria/submucosa with or without breaching the subepithelial basement membrane. Oral mucosal melanoma arising from melanocytes residing in the basal cell layer of the oral epithelium, on the other hand, give rise to three transient histopathological patterns: an epithelial *in-situ* pattern characterised by radial growth within the oral epithelium; an invasive pattern characterized by nodular aggregates of infiltrating epithelial melanoma cells within the lamina propria; and a combined form in which both junctional and nodular patterns occur [15, 55, 56]. Ultimately if not treated, all three patterns will become deeply invasive and metastatic.

Although the pathogenesis of oral mucosal melanoma is incompletely understood as yet, melanin, intermediates of melanin synthesis, and MC1R genetic polymorphism play roles in the pathogenesis of some cases of oral mucosal melanoma [57, 58]. Certain MC1R variants reduce the capacity for repair to damaged DNA [5], while the process of melanin biosynthesis, particularly of pheomelanin, may itself generate reactive oxygen species and other by-products that may be cytotoxic, genotoxic, or mutagenic causing DNA damage [4, 6].

It has been demonstrated that in some cases of oral mucosal melanoma, there is loss of the integrity of the melanosomal membrane with leakage into the cytoplasm and nucleoplasm of

http://dx.doi.org/10.5772/65567

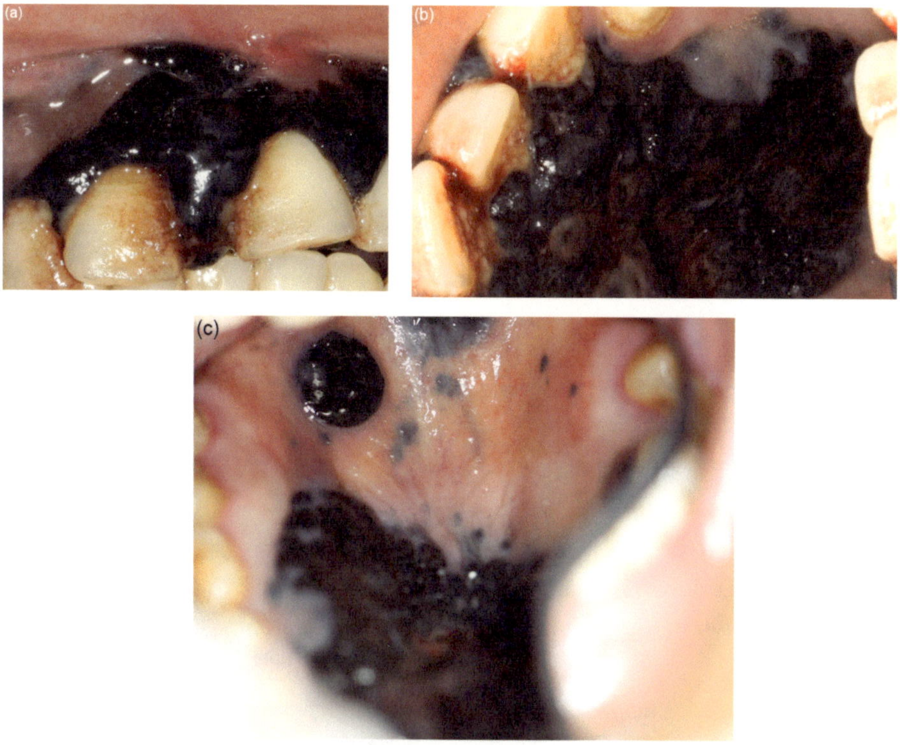

Figure 6. A 56-year-old black female with histologically confirmed melanoma of the hard palate and maxillary gingiva (a–c) of 3-year duration. It developed from a pigmented 'patch' of the buccal gingiva (reproduced with permission from Tlholoe et al. [15]).

the melanocytes, of toxic melanin particles, intermediates of melanogenesis and reactive oxygen species, with consequent DNA damage and increase in the risk of genetic mutations [59]. Furthermore, when leaked into the extracellular microenvironment, the intermediate metabolites of melanogenesis, which are also immunosuppressive, may promote evasion of immune responses by the initially transformed melanocyte precursors and their offspring melanoma cells, increasing the risk of melanomagenesis and melanoma growth [60].

Some oral mucosal melanomata are amelanotic, but most are heavily to very heavily melanin-pigmented. It is unknown whether the abnormally increased biosynthesis of melanin is the outcome of an early biopathological process in the development of oral mucosal melanoma, playing a role in the initial transformational events of melanocytes, or whether it is a manifestation of a malignant phenotype arising subsequently to the malignant transformation of normal melanocytes. As stated above, it has been reported that up to one-third of oral mucosal melanoma arise within fields of benign melanin hyperpigmentation [15, 47–49]. If this is true, then it is likely that either the upregulated process of melanin biosynthesis, the increased amount of melanin in the melanocytes, or both, occurring in the fields of benign oral melanin hyperpigmentation, constitute risk factors for transformation into oral mucosal melanomata.

Thus, in a nutshell, it is possible that the initial DNA damage that predisposes melanocytes to malignant transformation and cytogenetic alterations that later promote actual malignant transformation of the initially transformed melanocytes may be driven by dysregulation in melanin biosynthesis and by the increased cellular content of melanin [10].

Author details

Liviu Feller*, Razia A.G. Khammissa and Johan Lemmer

*Address all correspondence to: liviu.feller@smu.ac.za

Department of Periodontology and Oral Medicine, Sefako Makgatho Health Sciences University Pretoria, South Africa

References

[1] Barrett AW, Scully C. Human oral mucosal melanocytes: a review. *Journal of Oral Pathology & Medicine: Official Publication of the International Association of Oral Pathologists and the American Academy of Oral Pathology*. 1994;**23**:97–103.

[2] Amir E, Gorsky M, Buchner A, Sarnat H, Gat H. Physiologic pigmentation of the oral mucosa in Israeli children. *Oral Surgery, Oral Medicine, and Oral Pathology*. 1991;**71**:396–398.

[3] Squier CA, Kremer MJ. Biology of oral mucosa and esophagus. *Journal of the National Cancer Institute Monographs*. 2001;**29**:7–15.

[4] Feller L, Masilana A, Khammissa RA, Altini M, Jadwat Y, Lemmer J. Melanin: the biophysiology of oral melanocytes and physiological oral pigmentation. *Head & Face Medicine*. 2014;**10**:8.

[5] Lekalakala PT, Khammissa RA, Kramer B, Ayo-Yusuf OA, Lemmer J, Feller L. Oculocutaneous Albinism and squamous cell carcinoma of the skin of the head and neck in Sub-Saharan Africa. *Journal of Skin Cancer*. 2015;**2015**:167847.

[6] Feller L, Chandran R, Kramer B, Khammissa RA, Altini M, Lemmer J. Melanocyte biology and function with reference to oral melanin hyperpigmentation in HIV-seropositive subjects. *AIDS Research and Human Retroviruses*. 2014;**30**:837–843.

[7] Masilana A, Khammissa RAG, Lemmer J, Feller J. Oral medicine case book 66: Physiological/racial oral melanin hyperpigmentation. *SADJ: Journal of the South African Dental Association = tydskrif van die Suid-Afrikaanse Tandheelkundige Vereniging*. 2015;**70**:28–31.

[8] Mackintosh JA. The antimicrobial properties of melanocytes, melanosomes and melanin and the evolution of black skin. *Journal of Theoretical Biology*. 2001;**211**:101–113.

[9] Tobin DJ. Biochemistry of human skin--our brain on the outside. *Chemical Society Reviews*. 2006;**35**:52–67.

[10] Chandran R, Feller L, Lemmer J, Khammissa RA. HIV-associated oral mucosal melanin hyperpigmentation: A Clinical Study in a South African Population Sample. *AIDS Research and Treatment*. 2016;**2016**:8389214.

[11] Gondak RO, da Silva-Jorge R, Jorge J, Lopes MA, Vargas PA. Oral pigmented lesions: Clinicopathologic features and review of the literature. *Medicina Oral, Patologia Oral y Cirugia Bucal*. 2012;**17**:e919–e924.

[12] Yuan A, Woo SB. Adverse drug events in the oral cavity. *Oral Surgery, Oral Medicine, Oral Pathology and Oral Radiology*. 2015;**119**:35–47.

[13] Muller S. Melanin-associated pigmented lesions of the oral mucosa: presentation, differential diagnosis, and treatment. *Dermatologic Therapy*. 2010;**23**:220–229.

[14] Chandran R, Khammissa RAG, Lemmer J, Feller L. HIV associated oral melanin hyperpigmentation. *SADJ: Journal of the South African Dental Association = tydskrif van die Suid-Afrikaanse Tandheelkundige Vereniging*. 2014;**69**:370–371.

[15] Tlholoe MM, Khammissa RA, Bouckaert M, Altini M, Lemmer J, Feller L. Oral mucosal melanoma: some pathobiological considerations and an illustrative report of a case. *Head and Neck Pathology*. 2015;**9**:127–134.

[16] Nishimura EK, Jordan SA, Oshima H, Yoshida H, Osawa M, Moriyama M, et al. Dominant role of the niche in melanocyte stem-cell fate determination. *Nature*. 2002;**416**:854–860.

[17] Yamaguchi Y, Brenner M, Hearing VJ. The regulation of skin pigmentation. *The Journal of Biological Chemistry*. 2007;**282**:27557–27561.

[18] Barrett AW, Raja AM. The immunohistochemical identification of human oral mucosal melanocytes. *Archives of Oral Biology*. 1997;**42**:77–81.

[19] Squier CA, Waterhouse JP. The ultrastructure of the melanocyte in human gingival epithelium. *Archives in Oral Biology*. 1967;**12**:119–129.

[20] Hirobe T. Role of keratinocyte-derived factors involved in regulating the proliferation and differentiation of mammalian epidermal melanocytes. *Pigment Cell Research/ Sponsored by the European Society for Pigment Cell Research and the International Pigment Cell Society*. 2005;**18**:2–12.

[21] Hara M, Toyoda M, Yaar M, Bhawan J, Avila EM, Penner IR, et al. Innervation of melanocytes in human skin. *The Journal of Experimental Medicine*. 1996;**184**:1385–1395.

[22] Schiaffino MV. Signaling pathways in melanosome biogenesis and pathology. *The International Journal of Biochemistry & Cell Biology*. 2010;**42**:1094–1104.

[23] Slominski A, Tobin DJ, Shibahara S, Wortsman J. Melanin pigmentation in mammalian skin and its hormonal regulation. *Physiological Reviews*. 2004;**84**:1155–1228.

[24] Slominski A, Wortsman J, Paus R, Elias PM, Tobin DJ, Feingold KR. Skin as an endocrine organ: implications for its function. *Drug Discovery Today Disease Mechanisms*. 2008;**5**:137–144.

[25] Plonka PM, Passeron T, Brenner M, Tobin DJ, Shibahara S, Thomas A, et al. What are melanocytes really doing all day long…? *Experimental Dermatology*. 2009;**18**:799–819.

[26] Chi A, Valencia JC, Hu ZZ, Watabe H, Yamaguchi H, Mangini NJ, et al. Proteomic and bioinformatic characterization of the biogenesis and function of melanosomes. *Journal of Proteome Research*. 2006;**5**:3135–3144.

[27] Marks MS, Seabra MC. The melanosome: membrane dynamics in black and white. *Nature Reviews Molecular Cell Biology*. 2001;**2**:738–748.

[28] Tsukamoto K, Jackson IJ, Urabe K, Montague PM, Hearing VJ. A second tyrosinase-related protein, TRP-2, is a melanogenic enzyme termed DOPAchrome tautomerase. *The EMBO Journal*. 1992;**11**:519–526.

[29] Kauser S, Schallreuter KU, Thody AJ, Gummer C, Tobin DJ. Regulation of human epidermal melanocyte biology by beta-endorphin. *The Journal of Investigative Dermatology*. 2003;**120**:1073–1080.

[30] Grando SA, Pittelkow MR, Schallreuter KU. Adrenergic and cholinergic control in the biology of epidermis: physiological and clinical significance. *The Journal of Investigative Dermatology*. 2006;**126**:1948–1965.

[31] Sowden HM, Naseem KM, Tobin DJ. Differential expression of nitric oxide synthases in human scalp epidermal and hair follicle pigmentary units: implications for regulation of melanogenesis. *The British Journal of Dermatology*. 2005;**153**:301–309.

[32] Starner RJ, McClelland L, Abdel-Malek Z, Fricke A, Scott G. PGE(2) is a UVR-inducible autocrine factor for human melanocytes that stimulates tyrosinase activation. *Experimental Dermatology*. 2010;**19**:682–684.

[33] Buchner A, Merrell PW, Carpenter WM. Relative frequency of solitary melanocytic lesions of the oral mucosa. *Journal of Oral Pathology & Medicine: official publication of the International Association of Oral Pathologists and the American Academy of Oral Pathology*. 2004;**33**:550–557.

[34] Neville BW, Damm DD, Allen AM, Chi AC. Epithelial pathology. In: Falk K, editor. Oral and Maxillofacial Pathology. 1. St Louis, Missouri, USA: Elsevier; 2016. p. 350–355.

[35] Dummett CO. Physiologic pigmentation of the oral and cutaneous tissues in the Negro. *Journal of Dental Research*. 1946;**25**:421–432.

[36] Meleti M, Vescovi P, Mooi WJ, van der Waal I. Pigmented lesions of the oral mucosa and perioral tissues: a flow-chart for the diagnosis and some recommendations for the management. *Oral Surgery, Oral Medicine, Oral Pathology, Oral Radiology, and Endodontics*. 2008;**105**:606–616.

[37] Mergoni G, Ergun S, Vescovi P, Mete O, Tanyeri H, Meleti M. Oral postinflammatory pigmentation: an analysis of 7 cases. *Medicina Oral, Patologia Oral y Cirugia Bucal*. 2011;**16**:e11–e14.

[38] Li CC, Malik SM, Blaeser BF, Dehni WJ, Kabani SP, Boyle N, et al. Mucosal pigmentation caused by imatinib: report of three cases. *Head and Neck Pathology*. 2012;**6**:290–295.

[39] Nikitakis NG, Koumaki D. Laugier-Hunziker syndrome: case report and review of the literature. *Oral Surgery, Oral Medicine, Oral Pathology and Oral Radiology*. 2013;**116**:e52–e58.

[40] Fernandes D, Ferrisse TM, Navarro CM, Massucato EM, Onofre MA, Bufalino A. Pigmented lesions on the mucosa: a wide range of diagnoses. *Oral Surgery, Oral Medicine, Oral Pathology and Oral Radiology*. 2015;**119**:374–378.

[41] Kauzman A, Pavone M, Blanas N, Bradley G. Pigmented lesions of the oral cavity: review, differential diagnosis, and case presentations. *Journal (Canadian Dental Association)*. 2004;**70**:682–683.

[42] Grichnik JM, Rhodes AR, Sober AJ. Beign neoplasias and hyperplasias of melanocytes. In: Sydor AM, Pancotti R, editors. Fitzpatrick's Dermatology in General Medicine. 2: New York: The McGraw-Hill Companies, Inc; 2016. p. 1377–1416.

[43] Biesbrock AR, Aguirre A. Multiple focal pigmented lesions in the maxillary tuberosity and hard palate: a unique display of intraoral junctional nevi. *Journal of Periodontology*. 1992;**63**:718–721.

[44] Feller L, Masipa JN, Wood NH, Khamissa RA, Meyerov R, Lemmer J. Primary oral melanoma associated with HIV infection. *SADJ: Journal of the South African Dental Association = tydskrif van die Suid-Afrikaanse Tandheelkundige Vereniging*. 2008;**63**:016–017.

[45] Postow MA, Hamid O, Carvajal RD. Mucosal melanoma: pathogenesis, clinical behavior, and management. *Current Oncology Reports*. 2012;**14**:441–448.

[46] Kerr EH, Hameed O, Lewis JS, Jr., Bartolucci AA, Wang D, Said-Al-Naief N. Head and neck mucosal malignant melanoma: clinicopathologic correlation with contemporary review of prognostic indicators. *International Journal of Surgical Pathology*. 2012;**20**:37–46.

[47] Garzino-Demo P, Fasolis M, Maggiore GM, Pagano M, Berrone S. Oral mucosal melanoma: a series of case reports. *Journal of Cranio-Maxillo-Facial Surgery: Official Publication of the European Association for Cranio-Maxillo-Facial Surgery*. 2004;**32**:251–257.

[48] Kahn MA, Weathers DR, Hoffman JG. Transformation of a benign oral pigmentation to primary oral melanoma. *Oral Surgery, Oral Medicine, Oral Pathology, Oral Radiology, and Endodontics*. 2005;**100**:454–459.

[49] Shen ZY, Liu W, Bao ZX, Zhou ZT, Wang LZ. Oral melanotic macule and primary oral malignant melanoma: epidemiology, location involved, and clinical implications. *Oral Surgery, Oral Medicine, Oral Pathology, Oral Radiology, and Endodontics*. 2011;**112**:e21–e25.

[50] Benoist LB, van Looij MA. Images in clinical medicine. Melanoma of the oral cavity. *The New England Journal of Medicine*. 2013;**368**:e14.

[51] Martinelli-Klay CP, Laporte ML, Martinelli CR, Martinelli C, Lombardi T. Oral malignant melanoma initially misdiagnosed as a racial pigmentation: a case report. *Dermatopathology*. 2016;**3**:1–7.

[52] Carlson JA, Murphy M, Slominski A, Vincent WL. Evidence of skin field cancerization. In: Dakubo GD E, editor. Filed Cancerization: Basic Science and Clinical Applications. Ontario, Canada: Nova Science Publishers; 2011. p. 317–370.

[53] Wang X, Wu HM, Ren GX, Tang J, Guo W. Primary oral mucosal melanoma: advocate a wait-and-see policy in the clinically N0 patient. *Journal of Oral and Maxillofacial Surgery* 2012;**70**:1192–1198.

[54] Grichnik JM. Melanoma, nevogenesis, and stem cell biology. *The Journal of Investigative Dermatology*. 2008;**128**:2365–2380.

[55] Mohan M, Sukhadia VY, Pai D, Bhat S. Oral malignant melanoma: systematic review of literature and report of two cases. *Oral Surgery, Oral Medicine, Oral Pathology and Oral Radiology*. 2013;**116**:e247–e254.

[56] Barker BF, Carpenter WM, Daniels TE, Kahn MA, Leider AS, Lozada-Nur F, et al. Oral mucosal melanomas: the WESTOP Banff workshop proceedings. Western Society of Teachers of Oral Pathology. *Oral Surgery, Oral Medicine, Oral Pathology, Oral Radiology, and Endodontics*. 1997;**83**:672–679.

[57] Nan H, Kraft P, Hunter DJ, Han J. Genetic variants in pigmentation genes, pigmentary phenotypes, and risk of skin cancer in Caucasians. *International Journal of Cancer* 2009;**125**:909–917.

[58] Han J, Kraft P, Colditz GA, Wong J, Hunter DJ. Melanocortin 1 receptor variants and skin cancer risk. *International Journal of Cancer* 2006;**119**:1976–1984.

[59] Gidanian S, Mentelle M, Meyskens FL, Jr., Farmer PJ. Melanosomal damage in normal human melanocytes induced by UVB and metal uptake—a basis for the pro-oxidant state of melanoma. *Photochemistry and Photobiology*. 2008;**84**:556–564.

[60] Slavin S. Effector cells of experimental and clinical cellular adoptive immunobiology. In: Morstyn G, Sheridan W, editors. Cell Therapy: Stem Cell Transplantation, Gene Therapy, and Cellular Immunotherapy. Los Angeles: Cambridge University Press; 1996. p. 18–42.

Chapter 3

The Potential Role of the MCHR1 in Diagnostic Imaging: Facts and Trends

Cécile Philippe and Markus Mitterhauser

Additional information is available at the end of the chapter

http://dx.doi.org/10.5772/67185

Abstract

The neuropeptide melanin-concentrating hormone (MCH) plays a key role in energy maintenance by decreasing energy expenditure and stimulating feeding behavior. Furthermore, it is involved in diabetes, gut inflammation, sleep, depression, and cilia beat function. The biological function of MCH is mediated by two G-protein coupled receptors, MCH receptor 1 and 2 (MCHR1 and MCHR2). Since only the MCHR1 is functional in rodents, the physiological importance of MCHR2 remains unknown due to the lack of appropriate animal models. The involvement of the MCHergic system in a variety of pathologies, especially endocrinological diseases, such as obesity and diabetes, makes it interesting as a new target to treat human disorders. Many pharmaceutical companies have pursued the development of MCHR1 antagonists for the treatment of obesity. Moreover, positron emission tomography (PET) tracers targeting the MCHR1 have been developed in order to gain a deeper understanding of the role and distribution of the MCHR1. As a high-end technique, PET allows noninvasive in vivo visualization and quantification of receptor systems, as well as monitoring and following hormone receptor status and related pathologies. Therefore, a MCHR1 PET tracer could help to guide pharmacological intervention via the MCHR1.

Keywords: MCHR1, PET, imaging, tracer, antagonist

1. Introduction

Melanin-concentrating hormone (MCH) was first referenced as "melanophore-concentrating hormone" in studies examining the possible origin of a factor leading to lightning of fish skin color in 1955 [1]. Although speculations regarding the existence of such a factor date back to the 1930s—when studies of pigmentation changes in amphibians were performed [2]—the isolation of MCH succeeded only in 1983 [3]. It was gained from the pituitary gland of the salmon and characterized as a cyclic 17-amino acid polypeptide with a cysteine-cysteine disulfide bond [3]. In teleost fish, MCH is synthesized as a preprohormone in the pituitary gland and is secreted into the circulation, where it acts as the opponent of melanocyte-stimulating hormone (α-MSH) by lightning the skin color in response of the environmental background. This effect is due to the aggregation of pigment granules within melanophores [4, 5].

Several years after the discovery in fish, MCH was identified in mammalian brain, too. It was first isolated from rat hypothalamic fragments and characterized as a 19-amino acid peptide [6], which was found to be identical to human MCH [7] (structure shown in **Figure 1**). Generally, MCH is present in all groups of vertebrates from lampreys to humans and exhibits considerable structural coherence among different species [8]. MCH is derived by posttranslational cleavage from the C terminus of a 165 amino-acid precursor, pre-proMCH (ppMCH). The preprohormone is also generating two additional peptides, neuropeptide-glutamic acid-isoleucine (NEI), and neuropeptide-glycine-glutamic acid (NGE). The gene encoding MCH is called Pmch [7].

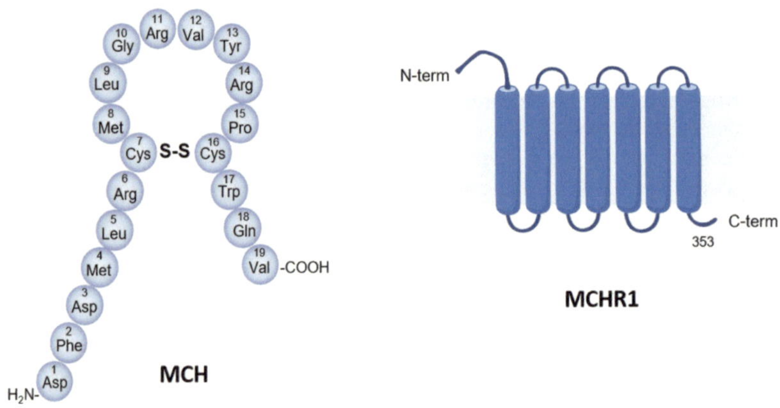

Figure 1. Structure of MCH and MCHR1.

In mammals, MCH is predominantly expressed in the lateral hypothalamus and zona incerta and projects broadly throughout the central nervous system (CNS) [9, 10]. Furthermore, it is also found in peripheral tissues, such as colonic epithelial cells [11], adipocytes [12], or beta-cells of the islets of Langerhans [13]. Although it is expressed in human melanocytes too, it has not been shown to affect human pigmentation [14]. In mammals, MCH operates as a neuropeptide, playing a key role in energy maintenance by decreasing energy expenditure and stimulating feeding behavior [15–17].

Initial efforts to identify a MCH receptor were based on binding assays by using radiolabeled MCH. MCH-binding sites were detected in a variety of cells and in rat brain [18, 19]. However, the existence of the MCH receptor remained obscure until 1999, when it was found to be identical to the previously cloned orphan G protein-coupled receptor somatostatin-like clone 1 (SLC-1), which exhibited about 40% homology to the five somatostatin receptors in its hydrophobic domains. The first MCH receptor (MCHR1) was identified simultaneously by several groups using different pharmacological approaches [20–24]. MCHR1 is 353 amino acids long and has all hallmark features of the G protein-coupled receptors, including seven

http://dx.doi.org/10.5772/67185

	Rat		Human	
Brain region				
Hypothalamus	++	[23]	+++	[29]
Thalamus	+++	[30]	++	[29]
Hippocampus	+++	[23]	++	[31]
Amygdala	++	[23]	+	[31]
Pons	+	[22]	++	[29]
Medulla oblongata	?		++	[29]
Substantia nigra	++	[23]	++	[32]
Nucleus accumbens	+++	[23]	++	[33]
Locus coeruleus	++	[23]	?	
Olfactory system	+++	[23]	?	
Cerebral cortex	++	[22]	+	[29]
Cerebellum	++	[34]	+	[29]
Peripheral tissues				
Pituarity	+	[23]	++	[33]
Eye	++	[23]	?	
Tongue	+	[23]	?	
Sceletal muscle	++	[23]	–	[32]
Adipose tissue	+	[23]	+	[33]
Liver	–	[23]	+	[32]
Heart	–	[23]	–	[32]
Thymus	?		–	[32]
Spleen	?		–	[32]
Kidney	–	[23]	–	[32]
Adrenal	+	[35]	+	[31]
Testis	+	[36]	+	[32]
Ovary	++	[36]	++	[32]
Placenta	?		–	[32]
Prostate	?		–	[32]
Stomach	?		–	[33]
Pancreas	+	[37]	+	[13]
Colonic epithelial cells	?		+	[11]

+++, high abundance; ++, moderate abundance; +, low abundance; –, not detectable; ?, not investigated.

Table 1. Binding sites of MCH/MCHR1 mRNA distribution pattern.

transmembrane helices, a DRY motif at the end of the third intracellular loop, and three potential glycosylation sites at the N-terminus [25] (structure shown in **Figure 1**).

Comparison of the human and the rodent receptors shows a high degree of conservation between species (human-rat 96% identity; human-mouse 95% identity), which is not unexpected knowing that the ligand structure is identical in human, rat, and mouse [25, 26].

Activation of MCHR1, which couples to multiple G proteins (G_i, G_o, and G_q), leads to an increase in intracellular Ca^{2+} accumulation via stimulation of phospholipase C, lowered cyclic adenosine monophosphate levels (cAMP) via inhibition of adenylate cyclase, and stimulation of extracellular-signal-regulated kinase [27]. In the CNS, activation of these signaling pathways has diverse effects, ranging from changes in gene expression to modulation of ion channel activity [28].

MCHR1 is predominately expressed in the brain; besides, it is also found in moderate to weak concentration in other tissues. An overview of the widespread MCHR1 distribution is given in **Table 1**.

In 2001, a second MCH receptor (MCHR2) was identified with 340 amino acids of length [32, 33, 38, 39]. The overall homology between the two MCH receptors is quite low; they share only 38% identical amino acids. It also seems that the signal transduction mechanism of MCHR2 is limited to the G_q protein, resulting in increased intracellular Ca^{2+} levels. Its expression profile is similar to MCHR1, with the highest expression in the brain, notably in the frontal cortex, amygdala, hippocampus, nucleus accumbens, and putamen [38]. Low levels were found in the thalamus, hypothalamus, medulla oblongata, and no expression was found in the cerebellum. Peripheral expression was shown in adipocytes, pancreas, prostate, and intestine [40]. MCHR2 is found to be a pseudogene in rodent species, but—interestingly—is known to be functional only in dogs, ferrets, rhesus monkeys, and humans [41]. Due to the lack of appropriate animal models, the physiological importance of MCHR2 remains unknown until now.

2. Physiological function of the MCHergic system

MCH plays a major role in energy homeostasis, e.g., the control of food intake, body weight, and metabolism [15–17]. It has been shown that MCH-deficient (MCH-KO) mice are leaner than wild-type mice and their body weight deficit is about 25% at 4 months of age. The decrease in body weight is the result of reduced food intake (hypophagia) and a slight increase in energy expenditure. Leptin levels are low, as would be expected in a lean mouse model. Otherwise, the mice appear to be normal, with normal levels of activity and normal fertility [42].

In contrast, MCH overexpressing (MCH-OE) mice developed—when placed on a high-fat diet—excessive obesity compared with wild-type mice. The reason, therefore, is that MCH-OE mice are hyperphagic. Furthermore, these animals had elevated blood glucose levels, significant hyperinsulinemia, and islet cell hyperplasia [43]. Chronic infusion of MCH in mice reproduced the hyperphagic obese phenotype seen in the transgenic mice (MCH-OE mice) [17].

Considering the receptor, MCHR1 knockout (MCHR1-KO) mice are—when maintained at regular feed—lean and have a reduced fat mass. The leanness is due to hyperactivity, increased energy expenditure, and altered metabolism. A dramatic 250% increase in running-wheel activity was reported [44]. Paradoxically, they were found to be hyperphagic compared with wild-type mice. This effect was interpreted as a compensatory response. When placed on a high-fat diet, they gain significantly less weight and are less susceptible to diet-induced obesity (DIO). Chronic infusion of MCH did not show any effects in MCHR1-KO mice [16].

Furthermore, MCH plays an important role in mediating the effects of leptin on energy homeostasis. Leptin is predominantly secreted from adipocytes and is referred as "adiposity signal," circulating in proportion to body-fat content [45]. Leptin-deficient (ob/ob) mice are obese, hyperphagic, insulin resistant, and had significantly increased MCH expression [15].

Crossing the ob/ob mice to the MCH-KO mice caused a significant reduction in body weight with increased energy expenditure, thermogenesis, and locomotor activity. Further, they showed improved glucose homeostasis [46].

In MCHR1-KO mice, which were crossed with the ob/ob mice, body-fat mass decreased and locomotor activity increased. No differences in body weight, food intake, or energy expenditure could be observed compared with the ob/ob mice. Despite being obese, MCHR1-KO ob/ob mice had improved insulin sensitivity [47].

All these findings illustrate the central role of MCH in regulating energy homeostasis: MCH promotes the conservation of body energy and the loss of MCH function leads to leanness. Conservation of energy is also one of the main functions of sleep. Hence, it seems likely that MCH has also a function in sleep regulation. Several research groups described the role of MCH as a sleep promoter (reviewed by Torterolo et al. [48]). MCHergic neurons project throughout the central nervous system, including areas involved in the control of sleep, such as the dorsal and median raphe nuclei. The fact the MCHergic neurons are active during sleep, especially during rapid eye movement (REM) sleep, indicates the MCH plays a critical role in the generation and maintenance of sleep. An increase in the duration of REM sleep is considered as an important biological marker for depression [49]. Furthermore, the expression of MCHR1 in serotonergic neurons, as well as MCHergic projections toward the limbic system, suggest a relevant role of MCH in the control of emotional states [50]. Several preclinical studies demonstrated the antidepressant and anxiolytic effects of MCHR1 antagonists (i.e. [51, 52]).

Furthermore, it has been shown that MCH contributes to maintain cerebrospinal fluid (CSF) homeostasis: MCHR1 is expressed in the ependymal cells of the third ventricle, where it is involved in the regulation of cilia beat frequency [53, 54]. This beating facilitates cerebrospinal fluid (CSF) circulation, which is crucial for brain function, as defects in ventricular cilia result in hydrocephalus. A lack of MCH receptor provokes a ventricular size increase as observed in MCHR1-KO mice. Brain penetrating MCHR1 antagonists may thus alter CSF flow, limiting their potential use as therapeutic agents.

Since both, MCH and MCHR1 are expressed in islets and clonal beta-cell lines [13], some peripheral effects of MCH were also observed. MCH has a direct effect on islet signaling pathways, insulin secretion, and insulin sensitivity [13]. As mentioned above, MCH-OE mice have substantial hyperinsulinemia and islet hyperplasia that is out of proportion with their degree of obesity [43]. In contrast, MCH-KO mice have normal or improved glucose tolerance, despite having less insulin release in response to a glucose load [55]. Furthermore, MCH stimulates leptin production in adipocytes, where the MCHR1 is present too [12]. MCH was also found to be a mediator of intestinal inflammation. MCH and MCHR1 mRNA expression are increased in human colitis. MCH-KO mice had a significant protection from induced colitis, suggesting that MCH has a proinflammatory role in the development of colitis [11]. Pmch and MCHR1 were also found in human immune cells. This may provide a link between allergic inflammation, asthma, and obesity [56, 57].

An overview of the multiple involvement of the MCHergic system is given in **Figure 2**.

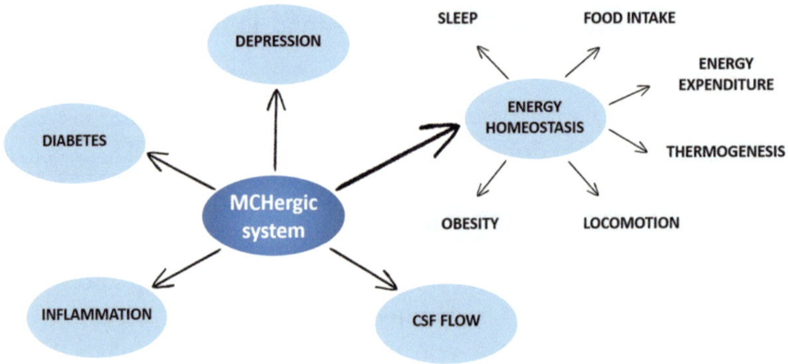

Figure 2. Involvement of the MCHergic system.

3. Targeting the MCHR1

The involvement of the MCHergic system in a variety of pathologies, especially endocrinological diseases such as obesity and diabetes, makes it interesting as a new target to treat human disorders. Since it has been shown that MCHR1 antagonists reduce body weight in rodents (reviewed by MacNeil [58]), several MCHR1 antagonists were developed in the last 15 years; some of them have entered clinical trials for the treatment of obesity [59], while some are in discussion of becoming anti-diabetic drugs [60].

The first functional, competitive MCHR1 antagonist was the $_{D}$-Ala^{11} analog of hMCH [61]. Subsequently, a series of analogs with antagonist activity were generated, such as Ac-(Ava$^{9-10}$, Ava$^{14-15}$)-hMCH(6-16)-NH_2 (also known as PMC-3881-PI) [62]. However, peptide MCHR1 antagonists are not able to cross the blood-brain barrier (BBB) and interact with the MCHR1 in the CNS. Therefore, small molecule MCHR1 antagonists were developed. The first nonpeptide antagonist was T-226296 (Takeda), an orally active compound that demonstrated high affinity and selectivity to the MCHR1 (Ki = 5.5 nM) [63].

Screening of a G protein-coupled receptor-based compound library against the human MCHR1 in a functional assay measuring intracellular Ca^{2+} mobilization resulted in the

http://dx.doi.org/10.5772/67185

discovery of a second nonpeptide antagonist, SNAP-7941 ((+)-methyl (4S)-3-{[(3-{4-[3-(acetylamino)phenyl]-1-piperidinyl}propyl)amino]carbonyl}-4-(3,4-difluorophenyl)-6-(methoxymethyl)-2-oxo-1,2,3,4-tetrahydro-5-pyrimidine carboxylate hydrochloride) (Synaptic/Lundbeck, **Figure 3** (1)) [51]. SNAP-7941 has an excellent binding affinity (Kd = 0.18 nM) and selectivity (>1000-fold) to the MCHR1. In rat brain sections, binding of radiolabeled SNAP-7941 ([^{3}H]SNAP-7941) was detected in the cerebral cortex, olfactory tubercle, claustrum, piriform cortex, hippocampus, amygdala, caudate-putamen, accumbens nucleus, hypothalamus, dorsal raphe, and locus coerulus [51]. This distribution pattern parallels the widespread MCHR1 expression in the brain [30]. Further, systemic pretreatment with SNAP-7941 (10 mg/kg, intraperintoneal (i.p.)) inhibited the increase in food intake induced by intra-cerebroventricular (i.c.v.) injection of MCH (3 nmol). Rats treated with SNAP-7941 (10 mg/kg, i.p.) twice a day for seven days gained 26% less weight compared with their littermates. SNAP-7941 was also capable to decrease milk consumption in satiated rats: 13% less milk consumption in rats treated with 3 mg/kg (i.p.), 41% with 10 mg/kg (i.p.), and 59% with 30 mg/kg (i.p.) compared with vehicle-treated rats. These results suggest that SNAP-7941 acts as an anorectic agent. To rule out that this anorectic effect was due to malaise, a taste aversion study was performed, which confirmed that the anorectic effect of SNAP-7941 was not a result of malaise. In DIO rats, SNAP-7941 (10 mg/kg, i.p., twice daily over four weeks) produced a sustained and consistent decrease in food consumption and body weight. This effect was reversible insofar as two weeks after the termination of the treatment, DIO rats previously treated with SNAP-7941 showed an increase in body weight and food consumption. A toxic effect of SNAP-7941 as a causal role for reduction in food intake and weight gain was ruled out by blood tests of the hepatic and renal function of the drug treated rats [51]. All these findings render the MCHR1 a viable target for the treatment of obesity. Additionally to the reported anorectic effects, SNAP-7941 evinced antidepressant and anxiolytic properties. A single oral dose of SNAP-7941 (3, 10, and 30 mg/kg) significantly decreased the duration of immobility and increased the swimming time of rats in the forced-swim test. Unfamiliar male rats showed increased social interaction time after acute treatment with SNAP-7941. Finally, the high-frequency vocalizations emitted from guinea pig pups separated from their mothers were significantly reduced after treatment with SNAP-7941 [51]. These anxiolytic effects were not supported by another group [64].

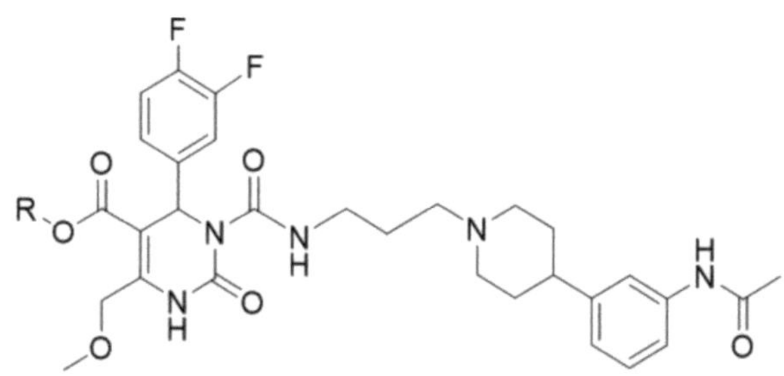

Figure 3. SNAP-7941 derivatives (**1**: SNAP-7941; **2**: [^{11}C]SNAP-7941; **3**: [^{18}F]FE@SNAP).

Several other small molecule antagonists were developed from pharmaceutical companies, e.g., ATC0065 and ATC0175 (Arena/Taisho collaboration) [65, 66], GW803430 (GlaxoSmithKline) [67], SNAP 94847 (Synaptic/Lundbeck) [52], AMG-076 (Amgen) [68], and NGD-4715 (Neurogen) [69]. All these drugs are BBB-penetrating compounds, which is mandatory to reduce body weight [70].

Due to medical need for oral anti-diabetics that give weight loss, a dual MCHR1 antagonist/dipeptidyl peptidase-4 (DPPIV) inhibitor had been developed recently [60].

Nevertheless, none of these molecules reached market authorization so far. Unfortunately, a significant number of MCHR1 antagonists showed a cardiovascular risk involving human ether-a-go-go gene (hERG) potassium channel inhibition and QT prolongation. hERG blockers are associated with lethal arrhythmias known as torsade de pointes. Structural requirements, such as a positively charged group and at least one distal aromatic/hydrophobic region, for MCHR1 potency correlate with hERG inhibition. Efforts in designing MCHR1 antagonists with improved selectivity over hERG are undertaken by several pharmaceutical companies; a few candidates have progressed to clinical development [71].

4. Diagnostic imaging

It has been described that changes of expression of the MCHR1 are related to various diseases, such as obesity, diabetes, gut inflammation, hydrocephalus, and depression. Thus, the MCHR1 seems to be a promising target for various clinical questions.

In the last decades, positron emission tomography (PET) has become a vital and versatile modality for modern medicine, as well as for applied medicinal research. At present, it belongs to the most sensitive methods for the purpose of molecular imaging [72]. However, the potential of nuclear molecular imaging and its capabilities strongly depends on the supply of potent and specific radiotracers for specific applications. Hence, the evaluation and development of new highly affine and selective radioligands is an indispensable part to increase the scope of application for molecular imaging.

A PET tracer for the MCHR1 comprises several advantages for clinicians and patients as the in vivo monitoring and following of the hormone receptor status and related pathologies. Moreover, it could support dose selection of MCHR1 antagonists in drug development [59] and would provide a deeper understanding of the involvement and distribution of the MCHR1. Therefore, specific MCHR1 imaging is of high clinical interest for status monitoring in endocrine pathologies such as obesity and diabetes.

As mentioned before, Borowsky et al. [51] presented the evaluation of the very potent MCHR1 antagonist SNAP-7941, which contains a methyl ester (**Figure 3**(1)), making it suitable for radiolabeling introducing either a [^{11}C]methyl moiety or a 2-[^{18}F]fluoroethyl moiety. On this basis, two potential PET tracers for the visualization of the MCHR1 were developed so far: [^{11}C]SNAP-7941 (**Figure 3**(2)) and [^{18}F]FE@SNAP (**Figure 3**(3)) [73–77].

The radiosynthesis of [^{11}C]SNAP-7941 could be conducted in a conventional synthesis module and consisted of a [^{11}C]methylation of the precursor SNAP acid. The optimal reaction conditions were found to be 2 min reaction time in acetonitrile at ≤25°C reaction temperature using 2 mg/ml precursor and [^{11}C]CH_3OTf as a methylation agent. Under these conditions, 2.9 ± 1.6 GBq (11.5 ± 6.4% at end of bombardment (EOB)) [^{11}C]SNAP-7941 could be produced. Radiochemical purity always exceeded 99% [73]. The specific radioactivity for the preclinical in vivo evaluation was 108.2 ± 56 GBq/μmol at the end of synthesis (EOS) [74]. The full radiosynthesis, purification procedure, and physiological formulation could be automated to guarantee a safe and reliable production. All quality control parameters were in accordance with the standards for human application [73]. In vitro binding assays on CHO-K1 cell membranes expressing the human MCHR1 and MCHR2, respectively, evinced high-binding affinity (Ki = 4.52 ± 0.7 nM for the MCHR1) and selectivity (Ki > 1000 nM for the MCHR2) of [^{11}C]SNAP-7941. The metabolic stability of [^{11}C]SNAP-7941 in plasma and against liver microsomes and carboxylesterases was very high: no degradation in human plasma over 60 min; <10% degradation by liver microsomes (human and rat) over 60 min and no decomposition by porcine carboxylesterases. Interestingly, in rat plasma, [^{11}C]SNAP-7941 was metabolized considerably: 50.4 ± 1% of intact compound were found after 60 min. The formation of a hydrophilic metabolite was observed. The plasma free fraction was found to be sufficient for imaging: f1 = 21.0 ± 1% in human plasma and even 32.4 ± 1% in rat plasma. The lipophilicity was in a range, where a passive diffusion through the BBB would be expectable: log $D^{7.4}$ = 3.29. However, in the in vivo experiments in healthy rats, [^{11}C]SNAP-7941 evinced to be a P-glycoprotein (P-gp) substrate, since after blocking with the P-gp inhibitor Tariquidar (TQD), the [^{11}C]SNAP-7941-uptake in the brain raised 3–5 times [74].

Of note, Borowsky et al. [51] observed a significant reduction in food intake and body weight in rats after i.p. application of SNAP-7941. Knowing that drugs have to cross the BBB to reduce body weight via central mechanisms [70], it is quite surprising that [^{11}C]SNAP-7941 evinced to be a P-gp substrate.

Nevertheless, no metabolite was found in the rat brains of the baseline and TQD-pretreatment group [74].

The biodistribution experiments showed both uptake in kidneys and in intestines. On the basis of the conducted experiments, a specification of the major way of excretion was not possible.

Uptake in MCHR1 expressing organs (adrenals, pancreas, eyes, muscle, and ovary) could be observed. Low uptake was found in peripheral fat tissue of the healthy rats [74].

Due to the advantage of the longer half-life of fluorine-18 (110 min) compared with carbon-11 (20 min), [^{18}F]FE@SNAP, the [^{18}F]fluoroethylated derivative of SNAP-7941 was developed as the second PET tracer for the MCHR1. Its radiosynthesis was challenging; no vessel-based approach succeeded. Only the preparation within a microfluidic device was successful, consisting of the direct [^{18}F]fluorination of the tosylated precursor (Tos@SNAP). Harsh reaction conditions had to be applied: high reaction temperature of 170°C and a short reaction time using an overall flow rate of 150 μl/min. Since the microfluidic device was not equipped for purification steps, the crude reaction mixture was purified in a conventional synthesis module. The radiosynthesis as well as the purification and physiological formulation were automated to guarantee safe and reliable production. The radiochemical yield ranged between 100 and 650 MBq, which was sufficient for every subsequent preclinical experiment. Specific radioactivity was 24.8 ± 12 GBq/μmol EOS. All quality control parameters were in accordance with the standards for human application [75].

[^{18}F]FE@SNAP evinced high-binding affinity and selectivity to the MCHR1: Kd = 2.9 ± 2.5 nM for the MCHR1; Ki > 1000 nM for the MCHR2. Similar to [^{11}C]SNAP-7941, it is highly stable in human plasma (only 3.9 ± 4% metabolism after 120 min), against liver microsomes (<10% metabolism after 60 min) and porcine carboxylesterases (Km = 347.3 μM) and rapidly metabolized in rat plasma (completely after 120 min). The plasma-free fraction was lower (f1 = 12.6 ± 0.2%) than for [^{11}C]SNAP-7941, but can still be considered high enough for imaging. Compared with SNAP-7941, the lipophilicity was slightly higher: $logD^{7.4}$ = 3.83 [76].

After IV administration of [^{18}F]FE@SNAP into healthy and conscious rats, ex vivo autoradiography of the brains evinced specific tracer uptake in the hypothalamic region and the ventricular system. In addition, preliminary small-animal PET measurements in healthy rats showed a high tracer uptake in the ventricular system (**Figure 4**) [77].

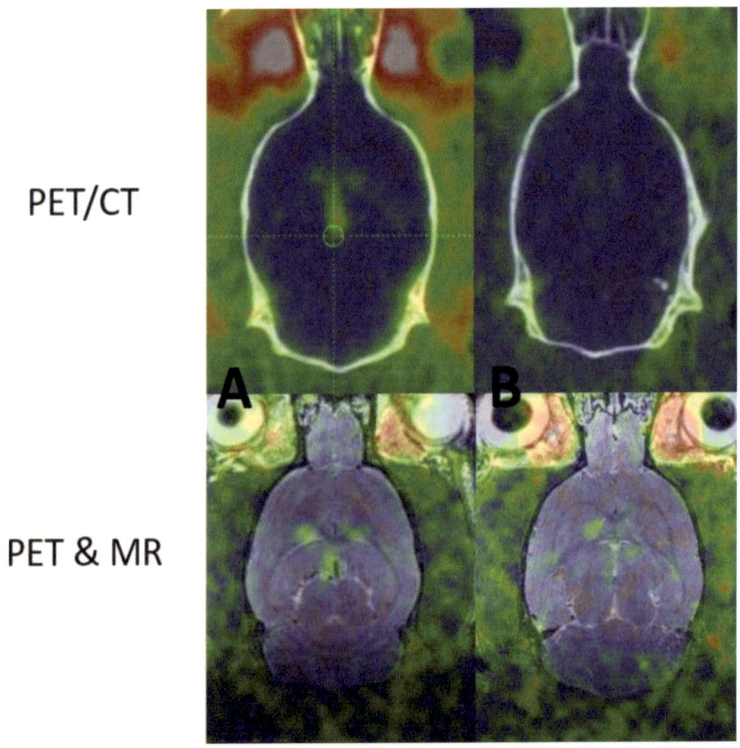

Figure 4. Exemplary small-animal PET/CT and PET and MR image of a rat brain with [^{18}F]FE@SNAP. Rats were anesthetized by isoflurane. 25 min after [^{18}F]FE@SNAP iv injection, vehicle (A) or 15mg/kg SNAP-7941 (B) were administered through the tail vein. 75 min after tracer injection, the rats were sacrificed. A significant reduction in tracer uptake could be observed after the administration of SNAP-7941.

Biodistribution experiments showed—similar to [^{11}C]SNAP-7941—uptake in kidneys as well as in the intestines. Uptake in MCHR1 expressing organs (adrenals, pancreas, eyes, muscle, and tongue) could be observed too [77].

So far, [^{11}C]SNAP-7941 and [^{18}F]FE@SNAP are the only described potential PET tracers for the MCHR1. Both proved high affinity and selectivity for the MCHR1; could be prepared in a reproducible and feasible way; evinced high metabolic stability in human plasma and against human and rat liver microsomes and showed specific uptake in preclinical in vivo experiments in healthy rats. Since the extent of MCHR1 expression in obese or diabetic rats is unknown, further experiments with [^{11}C]SNAP-7941 or [^{18}F]FE@SNAP in obese and diabetic rats will show the applicability of the MCHR1 concept. Furthermore, the quantitative analysis and evaluation of the preclinical data may provide high-impact predictive values upon further clinical application in PET imaging for diagnosis and further treatment of MCHR1-related diseases. A more detailed insight and better understanding of the involvement of the MCHR1 in various neurotransmitter systems can be obtained, which potentially provides the basis for the development of novel pharmaceuticals and eventually, may serve as new working standard for the evaluation of the MCHR1 as a potential binding site of atypical antipsychotic drugs, using PET.

5. Conclusion

MCH was initially characterized as a factor mediating color change in teleost fish. In response to the environmental background, MCH causes a concentration of pigment in melanophores, thus producing a lighting of pigmentation and has the opposite effect of α-MSH, which causes darkening of fish pigmentation. This phenomenon is crucial for fish to camouflage themselves from predators.

Although MCH is expressed in human melanocytes too, it has not been shown to affect human pigmentation so far. In mammals, MCH operates as a neuropeptide and involvement of MCH/MCHR1 was reported in obesity and diabetes, and MCHR1 has also been related to depression, colitis, and hydrocephalus. Therefore, MCHR1 imaging is of high clinical interest for monitoring these pathologies and guide pharmacological interventions. Specific MCHR1 PET tracers could serve as a useful tool for diagnostic imaging. So far, the first two PET tracer for the MCHR1, [^{11}C]SNAP-7941, and [^{18}F]FE@SNAP are evaluated in preclinical settings.

Author details

Cécile Philippe[1*] and Markus Mitterhauser[1, 2]

*Address all correspondence to: cecile.philippe@meduniwien.ac.at

1 Department of Biomedical Imaging and Image-guided Therapy, Division of Nuclear Medicine, Radiopharmacy and Experimental Nuclear Medicine, Medical University of Vienna, Austria

2 Ludwig Boltzmann Institute for Applied Diagnostics, Vienna, Austria

References

[1] Enami M. Melanophore-concentrating hormone (MCH) of possible hypothalamic origin in the catfish, Parasilurus. Science. 1955;**121**:36–37.

[2] Hogben LT and Slome D. The pigmentary effector system VI. The dual character of endocrine coordination in amphibian colour change. Proc R Soc London Biol Sci. 1931;**108**:10–53.

[3] Kawauchi H, Kawazoe I, Tsubokawa M, Kishida M, Baker BI. Characterization of melanin-concentrating hormone in chum salmon pituitaries. Nature. 1983;**305**:321–323.

[4] Barber LD, Baker BI, Penny JC, Eberle AN. Melanin concentrating hormone inhibits the release of αMSH from teleost pituitary glands. Gen Comp Endocrinol. 1987;**65**:79–86.

[5] Rogers SL, Tint IS, Fanapour PC, Gelfand VI. Regulating bidirectional motility of melanophore pigment granules along microtubules in vitro. Proc Natl Acad Sci USA. 1997;**94**:3720–3725.

[6] Vaughan JM, Fischer WH, Hoeger C, Rivier J, Vale W. Characterization of melanin-concentrating hormone from rat hypothalamus. Endocrinology. 1989;**125**:1660–1665.

[7] Presse F, Nahon JL, Fischer WH, Vale W. Structure of the human melanin concentrating hormone mRNA. Mol Endocrinol. 1990;**4**:632–637.

[8] Griffond B and Baker BJ. Cell and molecular cell biology of melanin-concentrating hormone. Int Rev Cytol. 2002;**213**:233–277.

[9] Bittencourt JC, Presse F, Arias C, Peto C, Vaughan JM, Nahon JL, et al. The melanin-concentrating hormone system of the rat brain: an immuno- and hybridization histochemical characterization. J Comp Neurol. 1992;**319**:218–245.

[10] Casatti CA, Elias CF, Sita LV, Frigo L, Furlani VCG, Bauer JA, et al. Distribution of melanin-concentrating hormone neurons projecting to the medial mammillary nucleus. Neuroscience. 2002;**115**:899–915.

[11] Kokkotou E, Moss AC, Torres D, Karagiannides I, Cheifetz A, Liu S. Melanin-concentrating hormone as a mediator of intestinal inflammation. Proc Natl Acad Sci USA. 2008;**105**:10613–10618.

[12] Bradley RL, Kokkotou E, Maratos-Flier E, Cheatham B. Melanin-concentrating hormone regulates leptin synthesis and secretion in rat adipocytes. Diabetes. 2000;**49**:1073–1077.

[13] Pissios P, Ozcan U, Kokkotou E, Okada T, Liew CW, Liu S, et al. Melanin concentrating hormone is a novel regulator of islet function and growth. Diabetes. 2007;**56**:311–319.

[14] Saito Y, Wang Z, Hagino-Yamagishi K, Civelli O, Kawashima S, Maruyama K, et al. Endogenous melanin-concentrating hormone receptor SLC-1 in human melanoma SK-MEL-37 cells. Biochem Biophys Res Commun. 2001;**289**:44–50.

[15] Qu D, Ludwig DS, Gammeltoft S, Piper M, Pelleymounter MA, Cullen MJ, et al. A role for melanin-concentrating hormone in the central regulating of feeding behaviour. Nature. 1996;**380**:243–247.

[16] Marsh DJ, Weingarth DT, Novi DE, Chen HY, Turmbauer ME, Chen AS, et al. Melanin-concentrating hormone 1 receptor-deficient mice are lean, hyperactive, and hyperphagic and have altered metabolism. Proc Natl Acad Sci USA. 2002;**99**:3240–3245.

[17] Ito M, Gomori A, Ishihara A, Oda Z, Mashiko S, Yumoto M, et al. Characterization of MCH-mediated obesity in mice. Am J Physiol Endocrinol Metab. 2003;284:E940–E945.

[18] Drozdz R, Siegrist W, Baker BI, Chluba-de Tapia J, Eberle AN. Melanin-concentrating hormone binding to mouse melanoma cells in vitro. FEBS Lett. 1995;**359**:199–202.

[19] Drozdz R and Eberle AN. Binding sites for melanin-concentrating hormone (MCH) in brain synaptosomes and membranes from peripheral tissues identified with highly titrated MCH. J Recept Sigmal Transduct Res. 1995;**15**:487–502.

[20] Bachner D, Kreienkamp H, Weise C, Buck F, Richter D. Identification of melanin concentrating hormone (MCH) as the natural ligand for the orphan somatostatin-like receptor 1 (SLC-1). FEBS Lett. 1999;457:522–524.

[21] Chambers J, Ames RS, Bergsma D, Muir A, Fitzgerald LR, Hervieu G, et al. Melanin-concentrating hormone is the cognate ligand for the orphan G-protein-coupled receptor SLC-1. Nature. 1999;**400**:261–265.

[22] Lembo PM, Grazzini E, Cao J, Hubatsch DA, Pelletier M, Hoffert C, et al. The receptor for the orexigenic peptide melanin-concentrating hormone is a G-protein-coupled receptor. Nat Cell Biol. 1999;**1**:267–271.

[23] Saito Y, Nothacker HP, Wang Z, Lin SH, Leslie F, Civelli O. Molecular characterization of the melanin-concentrating-hormone receptor. Nature. 1999;**400**:265–269.

[24] Shimomura Y, Mori M, Sugo T, Ishibashi Y, Abe M, Kurokawa T, et al. Isolation and identification of melanin-concentrating hormone as the endogenous ligand of the SLC-1 receptor. Biochem Biophys Res Commun. 1999;**261**:622–626.

[25] Lakaye B, Minet A, Zorzi W, Grisar T. Cloning of the rat brain cDNA encoding for the SLC-1 G protein-coupled receptor reveals the presence of an intron in the gene. Biochim Biophys Acta. 1998;**1401**:216–220.

[26] Kokkotou E, Tritos NA, Mastaitis JW, Slieker L, Maratos-Flier E. Melanin-concentrating hormone receptor is a target of leptin action in the mouse brain. Endocrinology. 2001;**142**:680–686.

[27] Hawes BE, Kil E, Green B, O'Neill K, Fried S, Graziano MP. The melanin-concentrating hormone receptor couples to multiple G proteins to activate diverse intracellular signaling pathways. Endocrinology. 2000;**141**:4524–4532.

[28] Curtis J and Finkbeiner S. Sending signals from the synapse to the nucleus: possible roles for CaMK, Ras/ERK, and SAPK pathways in the regulation of synaptic plasticity and neuronal growth. J Neurisci Res. 1999;**56**:88–95.

[29] Sone M, Takahashi K, Murakami O, Totsune K, Arihara Z, Satoh R, et al. Binding sites for the melanin-concentrating hormone in the human brain. Peptides. 2000;**21**:245–250.

[30] Hervieu GJ, Cluderay JE, Harrison D, Meakin J, Maycox P, Nasir S, et al. The distribution of the mRNA and protein products of the melanin-concentrating hormone (MCH) receptor gene, slc-1, in the central nervous system of the rat. Eur J Neurosci. 2000;**12**:1194–1216.

[31] Takahashi K, Totsune K, Murakami O, Sone M, Satho F, Kitamuro T, et al. Expression of the melanin-concentrating hormone receptor messenger ribonucleic acid in tumor tissues of pheochromocytoma, ganglioneuroblastoma, and neuroblastoma. J Clin Endocrinol Metab. 2001;**86**:369–374.

[32] Mori M, Harada M, Terao Y, Sugo T, Watanabe T, Shimomura Y, et al. Cloning of a novel G protein-coupled receptor, SLT, a subtype of the melanin-concentrating hormone receptor. Biochem Biophys Res Commun. 2001;**283**:1013–1018.

[33] Hill J, Duckworth M, Murdock P, Rennie G, Sabido-David C, Ames RS, et al. Molecular cloning and functional characterization of MCH2, a novel human MCH receptor. J Biol Chem. 2001;**276**:20125–20129.

[34] Saito Y, Cheng M, Leslie FM, Civelli O. Expression of the melanin-concentrating hormone (MCH) receptor mRNA in the rat brain. J Comp Neurol. 2001;**435**:26–40.

[35] Saito Y, Nothacker HP, Civelli O. Melanin-concentrating hormone receptor: an orphan receptor fits the key. TEM. 2000;**11**:299–303.

[36] Suply T, Della Zuana O, Audinot V, Rodriguez M, Beauverger P, Duhault J, et al. SLC-1 receptor mediates effect of melanin-concentrating hormone on feeding behavior in rat: a structure-activity study. J Pharmacol Exp Ther. 2001;**299**:137–146.

[37] Tadayyon M, Welters HJ, Haynes AC, Cluderay JE, Hervieu G. Expression on melanin concentrating hormone receptors in insulin-producing cells: MCH stimulates insulin release in RINm5F and CRI-G1 cell-lines. Biochem Biophys Res Commun. 2000;**275**:709–712.

[38] Sailer AW, Sano H, Zeng Z, McDonald TP, Pan J, Pong SS, et al. Identification and characterization of a second melanin-concentrating hormone receptor, MCH-2R. Proc Natl Acad Sci USA. 2001;**98**:7564–7569.

[39] Wang S, Behan J, O'Neill K, Weig B, Fried S, Laz T, et al. Identification and pharmacological characterization of a novel human melanin-concentrating hormone receptor, MCH-R2. J Biol Chem. 2001;**276**:34664–34670.

[40] An S, Cutler G, Zhao JJ, Huang SG, Tian H, Li W, et al. Identification and characterization of a melanin-concentrating hormone receptor. Proc Natl Acad Sci USA. 2001;**98**:7576–7581.

[41] Tan CP, Sano H, Iwaasa H, Pan J, Sailer AW, Hreniuk DL, et al. Melanin-concentrating hormone receptor subtypes 1 and 2: species-specific gene expression. Genomics. 2002;**79**:785–792.

[42] Shimada M, Tritos NA, Lowell BB, Flier JS, Maratos-Flier E. Mice lacking melanin-concentrating hormone are hypophagic and lean. Nature. 1998;**396**:670–674.

[43] Ludwig DS, Tritos NA, Mastaitis JW, Kulkarni R, Kokkotou E, Elmquist J, et al. Melanin-concentrating hormone overexpression in transgenic mice leads to obesity and insulin resistance. J Clin Invest. 2001;**107**:379–386.

[44] Zhou D, Shen Z, Strack AM, Marsh DJ, Shearman LP. Enhanced running wheel activity of both Mch1r- and Pmch-deficient mice. Regul Pept. 2005;**124**:53–63.

[45] Considine RV, Sinha MK, Heiman ML, Kriauciunas A, Stephens TW, Nyce MR, et al. Serum immunoreactive-leptin concentrations in normal-weight and obese humans. N Engl J Med. 1996;**334**:292–295.

[46] Segal-Lieberman G, Bradley RL, Kokkotou E, Carlson M, Trombly DJ, et al. Melanin-concentrating hormone is a critical mediator of the leptin-deficient phenotype. Proc Natl Acad Sci USA. 2003;**100**:10085–10090.

[47] Bjursell M, Gerdin AK, Ploj K, Svensson D, Svensson L, Oscarsson J, et al. Melanin-concentrating hormone receptor 1 deficiency increases insulin sensitivity in obese leptin-deficient mice without affecting body weight. Diabetes. 2006;**55**:725–733.

[48] Torterolo P, Scorza C, Lagos P, Urbanavivius J, Benedetto L, Pascovich C, et al. Melanin-concentrating hormone (MCH): role in REM sleep and depression. Front Neurosci. 2015;**9**:475.

[49] Adrien J. Neurobiological bases for the relation between sleep and depression. Sleep Med Rev. 2002;**6**:341–351.

[50] Palagini L, Baglioni C, Ciapparelli A, Gemignani A, Riemann D. REM sleep dysregulation in depression: state of the art. Sleep Med Rev 2013;**17**:377-390. Eur J Pharmacol. 2002;**438**:129–135.

[51] Borowsky B, Durkin MM, Ogozalek K, Marzabadi MR, DeLeon J, Lagu B, et al. Antidepressant, anxiolytic and anorectic effects of a melanin-concentrating hormone-1 receptor antagonist. Nat Med. 2002;**8**:825–830.

[52] David DJ, Klemenhagen KC, Holick KA, Saxe MD, Mendez I, Santarelli DA, et al. Efficacy of the MCHR1 antagonist N-[3-(1-{[4-(3,4-difluorophenoxy)phenyl]methyl}[4-piperidyl))-4-methylphenyl]-2-methylpropanamide (SNAP 94847) in mouse models of anxiety and depression following acute and chronic administration is independent of hippocampal neurogenesis. J Pharmacol Exp Ther. 2007;**321**:237–248.

[53] Conductier G, Martin AO, Risold PY, Jego S, Lavoie R, Lafont C, et al. Control of ventricular ciliary beating by the melanin concentrating hormone-expressing neurons of the lateral hypothalamus; a functional survey. Front Endocrinol. 2013;**4**:182.

[54] Conductier G, Brau F, Viola A, Langlet F, Ramkumar N, Dehouck B, et al. Melanin-concentrating hormone regulates beat frequency of ependymal cilia and ventricular volume. Nat Neurosci. 2013;**16**:845–847.

[55] Jeon JY, Bradley RL, Kokkotou EG, Marino FE, Wang X, Pissios P, et al. MCH–/– mice are resistant to aging-associated increase in body weight and insulin resistance. Diabetes. 2006;**55**:428–434.

[56] Verlaet M, Adamantidis A, Coumans B, Chanas G, Zorzi W, Heinen E, et al. Human immune cells express ppMCH mRNA and functional MCHR1 receptor. FEBS Lett. 2002;**527**:205–210.

[57] Sandig H, McDonald J, Gilmour J, Arno M, Lee TH, Cousins DJ. Human Th2 cells selectively express the orexigenic peptide, pro-melanin-concentrating hormone. Proc Natl Acad Sci USA. 2007;**104**:12440–12444.

[58] MacNeil DJ. The role of melanin-concentrating hormone and its receptor in energy homeostasis. Front Endocrinol. 2013;**4**:49.

[59] Luthin DR. Anti-obesity effects of small molecule melanin-concentrating hormone receptor 1 (MCHR1) antagonists. Life Sci. 2007;**81**:423–440.

[60] Gattrell WT, Sambrook Smith CP, Smith AJ. An example of designed multiple ligands spanning protein classes: dual MCH-1R antagonists/DPPIV inhibitors. Bioorg Med Chem Lett. 2012;**22**:2464–2469.

[61] Macdonald D, Murgolo N, Zhang R, Durkin JP, Yao X, Strader CD, et al. Molecular characterization of the melanin-concentrating hormone/receptor complex: identification of critical residues involved in binding and activation. Mol Pharmacol. 2000;**58**:217–225.

[62] Bednarek MA, Hreniuk DL, Tan C, Palyha OC, MacNeil DJ, Van der Ploeg LH, et al. Synthesis and biological evaluation in vitro of selective, high affinity peptide antagonists of human melanin-concentrating hormone action at human melanin-concentrating hormone receptor 1. Biochemistry. 2002;**41**:6383–9630.

[63] Takekawa S, Asami A, Ishihara Y, Terauchi J, Kato K, Shimomura Y, et al. T-226296: a novel, orally active and selective melanin-concentrating hormone receptor antagonist. Eur J Pharmacol. 2002;**438**:129–135.

[64] Basso AM, Bratcher NA, Gallagher KB, Cowart MD, Zhao C, Sun M, et al. Lack of efficacy of melanin-concentrating hormone-1 receptor antagonists in models of depression and anxiety. Eur J Pharmacol. 2006;**540**:115–120.

[65] Kanuma K, Omodera K, Nishiguchi M, Funakoshi T, Chaki S, Semple G, et al. Discovery of 4-(dimethylamino)quinazolines as potent and selective antagonists for the melanin-concentrating hormone receptor 1. Bioorg Med Chem Lett. 2005;**15**:2565–2569.

[66] Kanuma K, Omodera K, Nishiguchi M, Funakoshi T, Chaki S, Semple G, et al. Lead optimization of 4-(dimethylamino)quinazolines, potent and selective antagonists for the melanin-concentrating hormone receptor 1. Bioorg Med Chem Lett. 2005;**15**:3853–3856.

[67] Carpenter AJ and Hertzog DL. Melanin concentrating hormone receptor antagonists as potential antiobesity agents. Expert Opin Ther Patents. 2002;**12**:1639–1646.

[68] Andersen D, Storz T, Liu P, Wang X, Li L, Fan P, et al. Stereoselective synthesis of a MCHr1 antagonist. J Org Chem. 2007;**72**:9648–9655.

[69] Rokosz LL. Discovery and development of melanin-concentrating hormone receptor 1 antagonists for the treatment of obesity. Expert Opin Drug Discovery. 2007;**2**:1301–1327.

[70] Hu XE, Wos JA, Dowty ME, Suchanek PM, Ji E, Chambers JB, et al. Small-molecule melanin-concentrating hormone-1 receptor antagonists require brain penetration for inhibition of food intake and reduction in body weight. J Pharmacol Exp Ther. 2008;**324**:206–213.

[71] Méndez-Andino JL and Wos JA. MCH-R1 antagonists: what is keeping most research programs away from the clinic? Drug Discovery Today. 2007;**12**:972–979.

[72] Das BK (ed.). Positron emission tomography. A Guide for Clinicians. Springer India 2015.

[73] Philippe C, Schirmer E, Mitterhauser M, Shanab K, Lanzenberger R, Karanikas G, et al. Radiosynthesis of [^{11}C]SNAP-7941—the first PET-tracer for the melanin concentrating hormone receptor 1 (MCHR1). Appl Rad Isot. 2012;**70**:2287–2294.

[74] Philippe C, Nics L, Zeilinger M, Kuntner C, Wanek T, Mairinger S, et al. Preclinical in vitro & in vivo evaluation of [^{11}C]SNAP-7941—the first PET tracer for the melanin concentrating hormone receptor 1. Nucl Med Biol. 2013; **40**:919–925.

[75] Philippe C, Ungersboeck J, Schirmer E, Zdravkovic M, Nics L, Zeilinger M, et al. [^{18}F]FE@SNAP—A new PET tracer for the melanin concentrating hormone receptor 1 (MCHR1): Microfluidic and vessel-based approaches. Bioorg Med Chem. 2012;**20**:5936–5940.

[76] Philippe C, Nics L, Zeilinger M, Schirmer E, Spreitzer H, Karanikas G, et al. Preparation and first preclinical evaluation of [^{18}F]FE@SNAP: a potential PET tracer for the melanin concentrating hormone receptor 1 (MCHR1). Sci Pharm. 2013;**81**:625–639.

[77] Philippe C, Haeusler D, Scherer T, Fürnsinn C, Zeilinger M, Wadsak W, et al. [^{18}F]FE@SNAP—a specific PET tracer for melanin-concentrating hormone receptor 1 imaging? EJNMMI Res. 2016;**6**:31.

Section 3

Fungal Melanin

Chapter 4

Production of Melanin Pigment by Fungi and Its Biotechnological Applications

Sandra R. Pombeiro-Sponchiado, Gabriela S. Sousa,
Jazmina C. R. Andrade, Helen F. Lisboa and
Rita C. R. Gonçalves

Additional information is available at the end of the chapter

http://dx.doi.org/10.5772/67375

Abstract

Production of the microbial pigments is one of the emerging fields of research due to a growing interest of the industry for safer products, easily degradable and eco-friendly. Fungi constitute a valuable source of pigments because they are capable of producing high yields of the substance in the cheap culture medium, making the bioprocess economically viable on the industrial scale. Some fungal species produce a dark-brown pigment, known as melanin, by oxidative polymerization of phenolic compounds, such as glutaminyl-3,4-dihydroxybenzene (GDHB) or catechol or 1,8-dihydroxynaphthalene (DHN) or 3,4-dihydroxyphenylalanine (DOPA). This pigment has been reported to act as "fungal armor" due to its ability to protect fungi from adverse conditions, neutralizing oxidants generated in response to stress. Apart from the scavenging activity, melanin exhibits other biological activities, including thermoregulatory, radio- and photoprotective, antimicrobial, antiviral, cytotoxic, anti-inflammatory, and immunomodulatory. Studies have shown that the media composition and cultivation conditions affect the pigment production in fungi and the manipulation of these parameters can result in an increase in pigment yield for large-scale pigment production. This chapter presents a comprehensive discussion of the research on fungal melanin, including the recently discovered biological activities and the potential use of this pigment for various biotechnological applications in the fields of biomedicine, dermocosmetics, materials science, and nanotechnology.

Keywords: fungi, pigment, melanin, biological activity, industrial applications

1. Introduction

Considering the harmful effects of synthetic dyes on human health and to the environment, developmental process for obtaining pigments from natural sources has become significant worldwide. Microbial pigments have gained attention owing to a growing interest of the industry in safer products, easily degradable, eco-friendly and do not cause harmful effects. The pigment production from microorganisms is considered more advantageous because it is a more efficient and cost-effective process than chemical synthesis of pigments. Microorganisms are also more feasible sources of pigments in comparison to pigments extracted from plants and animals because they do not have seasonal constraints, do not compete for limited farming land with actual foods, and can be produced easily in the cheap culture medium with high yields [1–6]. Besides, the microorganisms produce an extraordinary range of pigments that include several chemical classes such as carotenoids, melanins, flavins, phenazines, quinones, monascins, violacein, or indigo, as shown in **Table 1**.

Pigment	Microorganism
Indigoidine (blue-green)	*Streptomyces aureofaciens* CCM 323, *Corynebacterium insidiosum*
Carotenoid (orange)	*Gemmatimonas aurantiaca* T-27
Melanin (black-brown)	*Kluyveromyces marxianus, Streptomyces chibanensis, Cryptococcus neoformans, Aspergillus* sp., *Wangiella dermatitidis, Sporothrix schenckii,* and *Burkholderia cepacia*
Prodigiosin (red)	*Serratia marcescens, Rugamonas rubra, Streptoverticillium rsubrireticuli, Serratia rubidaea, Vibrio psychroerythrus, Alteromonas rubra,* and *Vibrio gazogenes*
Zeaxanthin (yellow)	*Staphylococcus aureus, Vibrio psychroerythrus, Streptomyces* sp., and *Hahella chejuensis*
Canthaxanthin (orange)	*Monascus roseus, Bradyrhizobium* sp.
Xanthomonadin (yellow)	*Xanthomonas oryzae*
Astaxanthin (red)	*Phaffia rhodozyma, Haematococcus pluvialis*
Violacein (purple)	*Janthinobacterium lividum*
Anthraquinone (red)	*Paecilomyces farinosus*
Halorhodopsin and rhodopsin (pink	*Halobacterium halobium*
Rosy pink	*Lamprocystis roseopersicina*
Violet/purple	*Thiocystis violacea, Thiodictyon elegans*
Rosy peach	*Thiocapsa roseopersicina*
Orange brown	*Allochromatium vinosum*
Pink/purple violet	*Allochromatium warmingii*

Table 1. Pigments produced by different microorganisms. Adapted from Ref. [3].

Among microbial species, fungi represent an economically significant source of these compounds because they can act as microbial cell factories producing high yields of metabolites with great diverse chemical structures combined with ease of large-scale cultivation [7–9].

As shown in **Table 1**, some fungal species produce a dark-brown pigment, known as melanin. In general, this pigment is located in the outermost layer of the cell wall associated with chitin (referred as cell wall-bound melanin), but in some fungi, melanin can also be found outside the fungal cell, usually in the form of granules in culture fluids [10].

Fungal melanins are negatively charged, hydrophobic pigments of high molecular weight formed by oxidative polymerization from phenolic and/or indolic compounds, such as glutaminyl-3,4-dihydroxybenzene (GDHB) or catechol or 1,8-dihydroxynaphthalene (DHN) or 3,4-dihydroxyphenylalanine (DOPA). Most Ascomycota fungi synthesize DHN-melanin from the polyketide synthase pathway, whereas few species are able to produce melanin through L-DOPA, in a pathway that resembles mammalian melanin biosynthesis [11–13].

The melanin pigment is not essential for fungal development, but it has been reported to act as "fungal armor" due to its ability to protect the microorganisms from harmful environmental conditions. In vitro studies have shown that melanized fungi are more resistant to UV light-induced and oxidant-mediated damages, temperature extremes, hydrolytic enzymes, heavy metal toxicity, and antimicrobial drugs than those nonmelanized [10, 14–17]. Recent studies have shown that in industrial and roadside areas, there is an increase in the proportion of dark melanin-containing fungi, as *Cladosporium* and *Alternaria*, which were more resistant to contamination by heavy metals and unsaturated hydrocarbons. Radionuclide contamination also led to a change in fungal communities, with an increased proportion of melanized fungi. For example, melanized fungal species as *Cladosporium* spp., *Alternaria alternata*, *Aureobasidium pullulans*, and *Hormoconis resinae* were found to colonize the walls of the damaged reactor at Chernobyl where they are exposed to a high constant radiation field [18, 19].

The presence of melanin in the cell wall is also correlated with enhanced virulence of parasitic fungi, as *Paracoccidioides brasiliensis*, *Sporothrix schenckii*, and *Exophiala* (*Wangiella*) *dermatitidis* [17, 20, 21]. This pigment protects the conidia against digestion by proteases and hydrolases secreted by competitive microorganisms or against bactericidal and fungicidal proteins of animal origin, such as defensins, magainins, or protegrins [22]. This effect was observed for *Cryptococcus neoformans*, whose in vitro melanization has been associated with resistance against host effector cells, oxidants, microbicidal peptides, and amphotericin B [23–25], and in *Wangiella* (*Exophiala*) *dermatitidis*, when the polyketide synthase gene WdPKS1 associated on melanin production was disrupted, this strain has become more susceptible to voriconazole and amphotericin B [26]. Others studies suggest that melanin contributes to fungal pathogenesis because this pigment alters the host defense response mechanisms, decreases phagocytosis, and reduces the toxicity of microbicidal peptides, reactive oxygen species, and antifungal drugs as well as to play a significant role in fungal cell wall mechanical strength [27, 28].

Although the molecular structure of fungal melanin remains enigmatic, significant progress has been made in understanding particular aspects of its macro- and microstructure. These advances allow to elucidate the molecular mechanisms of the various biological functions of melanin [22]. Studies have shown that the effect of melanin enhancing the survival of fungi under adverse conditions can be mainly due to its powerful free radical scavenger properties, acting as a "sponge" for other free radicals generated by the fungus in response to environmental stress [20, 29, 30]. Apart from this scavenging ability, melanin exhibits other biological activities, including thermoregulatory, photoprotective, antimicrobial, antiviral, cytotoxic, anti-inflammatory, radioprotective, and immunomodulatory [13, 17, 18, 31–34].

Since melanin has characteristics of functional materials and bioorganic, a growing number of researchers see this pigment with great interest, taking advantage of their properties for numerous biotechnological applications in cosmetics, pharmaceutical, electronic, and food processing industries [12, 19, 35].

The purpose of this chapter involves a comprehensive discussion of the research on fungal melanin, including the recently discovered biological activities and the potential use of this pigment for several biotechnological applications. Additionally, we discussed the ways to explore the metabolic potential of the pigment-producing fungi by manipulation of cultivation conditions to improve performance of the process, increasing yields, and reducing cost, for large-scale production.

2. Factors influencing the melanin production

Microbial pigment production is now one of the emerging fields of research due to its potential for various industrial applications, as foodstuff, cosmetics, pharmaceutical, and textile manufacturing processes. However, it is known that for the success of microbial fermentation processes, it is necessary to choose the correct productive culture strain and to determine the appropriate cultivation conditions [4, 8, 36].

An ideal pigment-producing microorganism should be capable of using a wide range of C and N sources; must be tolerant to pH, temperature, and minerals concentration; and must give reasonable pigments yield. The nontoxic and nonpathogenic natures, coupled with easy separation from cell biomass, are also preferred qualities. The potential of using filamentous fungi as pigment sources is due to their extraordinary metabolic versatility because they can be cultivated over a wide range of temperatures (10–50°C), pH (2–11), salinity (0–34%), and water activity (0.6–1) and under oligotrophic or nutrient-rich conditions. They can grow in different culture systems (submerged and solid), and fermentation protocols have been established for large-scale industrial processes. In addition, these organisms can be genetically modified to increase productivity and quality of the produced pigments [37, 38].

In order to improve performance and reduce the cost of pigments produced by microbial fermentation, it is essential to identify the nutritional and physical factors that have a greater influence on the cell growth and metabolite biosynthesis [4, 6, 39, 40].

Several studies have shown that the composition of the growth medium, nature and concentration of carbon and nitrogen sources, minerals, vitamins, temperature, pH, the presence of oxygen and aeration, light, stress, and irradiation, among others, affect the growth and pigment production in fungi and that the manipulation of the culture conditions can result in enhanced pigment production [41–47].

Experimental evidences indicate that the growth temperature influences the performance of the pigment production process, but this effect depends on the type of organism. *Pseudomonas* requires 35–36°C for its growth and pigment production, while in *Monascus purpureus*, maximum pigment production was observed at 30°C with a reduction of the yield at 37°C [48]. Another study in *Monascus* sp. *J101* reported that the yield of pigment at 25°C was ten times higher than at 30°C, probably due to long growing (120 hours) and lower viscosity of the broth at 25°C compared to 30°C [49]. Studies developed in our laboratory, using a melanin-overproducing mutant (MEL1) from *Aspergillus nidulans* fungus, showed that the higher production of pigment occurred at incubation temperature of 28°C compared to 37°C [50].

Researches support that the pH of the medium also affects the growth of fungi and type of pigment produced. In species of *Monascus*, the pH influences the yield and quality of the produced pigment, with the highest red pigment excretion and production at alkaline pH [51, 52]. Studies on wood-inhabiting fungi indicate that pH of the substrate potentially plays an important role in fungal melanin formation. Fungi *Trametes versicolor* and *Xylaria polymorpha* tested on wood substrates produced maximum pigmentation at the pH range 4.5–5.0, except for *Scytalidium cuboideum*, which produce maximum intensity of red pigment at pH 6 and blue pigment at pH 8 [53]. In our study with the hypermelanized mutant (MEL1) from *A. nidulans*, we observed an increase in the production of pigment when the initial pH of the culture was at 6.8 compared to pH 8.0 [50]. Metabolically, the effects of pH and temperature on fungal pigment production is associated with changes in protein activity, so that the culture conditions may control certain activities such as cell growth, production of primary and secondary metabolites, fermentation, and oxidation processes of the cell [54].

The influence of light on intra- and extracellular pigment production was studied in five pigment-producing fungi: *M. purpureus*, *Isaria farinosa*, *Emericella nidulans*, *Fusarium verticillioides*, and *Penicillium purpurogenum* [55]. These authors concluded that the cultivation in the total absence of light increased biomass and production of extracellular and intracellular pigments in all fungi. The fungi grown under red light have no effect, and green or yellow light resulted in worsening effect in all the fungi, thus postulating the existence of photoreceptors responsive to dark and light in all the fungi. In a similar study, [56] noted that the production of pigment by *Monascus* species also was favored when the fungus was grown in the dark.

Some studies report that the pigment synthesis requires proper aeration probably related to the oxygen dependency of some enzymatic reactions responsible for the production of pigment. In *Monascus ruber*, it was observed that the highest levels of pigments production were obtained at an aeration rate of 0.05 L min^{-1}, which appeared to be clearly sufficient for providing the fungus with oxygen and removing carbon dioxide [57]. In our studies, it was noted that no melanin pigment production takes place during stationary cultivation of hypermelanized mutant (MEL1) from *A. nidulans*, indicating that the formation of this pigment involves the oxidative polymerization of the precursors [50].

Carbon and nitrogen are necessary for cellular metabolism, and these sources are related to the formation of biomass, the type produced pigment, and the yield of the desired substance. These nutrients may regulate the expression of genes of interest and activate important metabolic pathways for the production of pigments [45, 58, 59]. In general, glucose, an excellent carbon source for growth, interferes with the formation of many secondary metabolites, including pigments. For example, the pigment production by *Penicillium* sp. was evaluated in the presence of 10 different carbon sources, and the maximum mycelial growth was obtained with fructose, whereas the maximum pigment production was obtained with soluble starch [60]. This result shows that the increased biomass does not necessarily result in increased pigment production because pigments produced by fungi are secondary metabolites whose production usually occurs at the late growth phase (idiophase) of these microorganisms [61]. The pigment production capability of fungal species belonging to the genera *Penicillium, Aspergillus, Epicoccum, Lecanicillium,* and *Fusarium* was evaluated in different culture media, and the results showed that the complex media, as potato dextrose (PD) and malt extract (ME), favored increased pigment production [47]. According to the authors, these media contain nutrients that can regulate the expression of genes of interest and activate metabolic pathways important for the production of pigments.

Studies have demonstrated that the promoting or repressing effect of a nitrogen source on pigment production is strain dependent. It has been reported that various types of peptone, used as a nitrogen source, are able to promote an increase in the production of pigments in many species of fungi [55, 59, 62, 63]. However, *M. purpureus* was not able to grow in media containing peptone, and a maximum yield of the pigment was achieved when the media were supplemented with yeast extract (1%) and monosodium glutamate (5%) as nitrogen source [41]. In *M. ruber*, the use of glutamic acid as a nitrogen source showed promising results, either as stimulating the accumulation of extracellular pigments or contributing to increase the efficiency of the pigment production process [45]. The production of high amounts of extracellular melanin by the fungus *Gliocephalotrichum simplex* was obtained in cultures supplemented with tyrosine (2.5%) and peptone (1%) [64].

The optimization of medium composition is an important strategy to increase pigment production because some sources of carbon and nitrogen can be more easily assimilated and promote higher yields of the desired product. During the optimization experiments to enhance the production of melanin by *Auricularia auricula*, it was observed that soluble starch, tyrosine, peptone, $CaCO_3$, and K_2HPO_4 had positive effects, while glucose, $(NH_4)_2SO_4$, $MgSO_4$, $CuSO_4$, and $FeSO_4$ negatively impacted melanin production [46]. In other study with *A. auricula*, it was observed that yeast extract, tyrosine, and lactose have significant effects on pigment production and the optimization of medium resulted in 2.14-fold higher melanin concentration than that of the unoptimized medium [65].

Since the substrates for the production of pigment strongly influence the cost of the bioprocess, there is a need to select cheap and efficient substrates to make the process economically viable on the industrial scale. Large amounts of agro-industrial residues generated from diverse economic activities have attracted strong industry interest on the utilization of these residues as inexpensive substrates to support the growth of microorganisms in bioprocesses.

This strategy may represent an added value to the industry and also helps in solving pollution problems, reducing or preventing their disposal in the environment [1, 66, 67].

Various studies have reported the successful utilization of agro-industrial residues for the production of fungal pigments. The use of corn cob powder as a substrate for production of pigments by *M. purpureus* resulted in greater pigment production [68] than other substrates, as jackfruit seed [69], corn steep liquor [70], and grape waste [71]. In the black yeast *Hortaea werneckii*, it was observed that rice bran acts as the cheapest source for increased production of melanin by than wheat bran and coconut cake [72]. Wheat bran extract, L-tyrosine, and $CuSO_4$ represent the best combination of medium components to obtain the maximum melanin yield from the fungus *A. auricula* in submerged culture [73]. A study conducted in our laboratory evaluated the use of corn steep liquor, sugarcane bagasse, and molasses as nutritional source on pigment production by melanin-overproducing mutant (MEL1) from *A. nidulans*. We observed that, in the presence of 0.2% corn steep liquor, an increase in the pigment production occurred, while a high yield of biomass was obtained at a concentration of 2%. The supplementation of medium with molasses and sugar cane bagasse hydrolysate did not have a positive effect on pigment production but promoted an increase in the fungal growth. These results indicate that corn steep liquor contains substances that stimulate the synthesis of pigment and it represents a low-cost fermentation medium for large-scale production of the pigment melanin by MEL1 mutant for future industrial applications [74].

3. Pathways of melanin biosynthesis

Various techniques, including electron paramagnetic resonance [75], X-ray diffraction [76], infrared, ultraviolet and visible spectroscopy [77], and nuclear magnetic resonance [78], have been used to elucidate the melanin structure from different organisms. These studies have shown that fungi can produce different types of melanins by oxidative polymerization of phenolic or indolic compounds [11, 27].

Melanin in cell walls of *Basidiomycotina* is derived from phenolic precursors, as glutaminyl-3,4-dihydroxybenzene (GDHB) or catechol. In the parasitic fungus *Ustilago maydis*, polymerization of catechol dimers with the formation of fibrils of melanin was shown [79]. The precursor of melanin in *Agaricus bisporus* and other *Basidiomycetes* is a metabolite of the shikimic acid pathway-γ-glutaminyl-4-hydroxybenzene oxidized under the action of peroxidase and/or phenolase into γ-glutaminyl-3,4-benzoquinone, followed by its polymerization [80]. *C. neoformans*, a pathogenic basidiomycetous yeast, is known to synthesize DOPA-melanin when o-diphenolic compounds, such as 3,4-dihydroxyphenylalanine, are present in the culture medium. This fungus may use a wide array of substrates, such as D- and L-dopamine [81], homogentisic acid [82], catecholamines, and other phenolic compounds [83], maximizing its ability to produce melanin. Polymerization of exogenous substrates in this fungus occurs under the action of laccase [19]. However, it is important to emphasize that different properties are observed for melanins derived from different substrates. Comparison of the catecholamines L-dopa, methyldopa, epinephrine, and norepinephrine shows differences in term of color, yield, and thickness of the cell wall melanin layer. It was also observed that the pigments vary in the strength of the stable free radical signal detectable by EPR [13, 83].

In the Ascomycota fungi, melanin pigment is generally synthesized from the pentaketide pathway in which 1,8-dihydroxynaphthalene (DHN) is the immediate precursor of the polymer, as described by Bell and Wheeler [11] based on genetic and biochemical evidence obtained from *Verticillium dahliae* and *W. dermatitidis* [84, 85]. **Figure 1** shows a general model for fungal dihydroxynaphthalene (DHN)-melanin biosynthesis. In this pathway, the polyketide synthase (PKS) converts malonyl-CoA to 1,3,6,8-tetrahydroxynaphthalene (1,3,6,8-THN), which undergoes several reduction and dehydration reactions to produce scytalone, 1,3,8-trihydroxynaphthalene (THN), and vermelone. A further dehydration step leads to the intermediate 1,8-dihydroxynaphthalene (DHN), which is polymerized to DHN-melanin, possibly by a laccase enzyme [10, 13, 27].

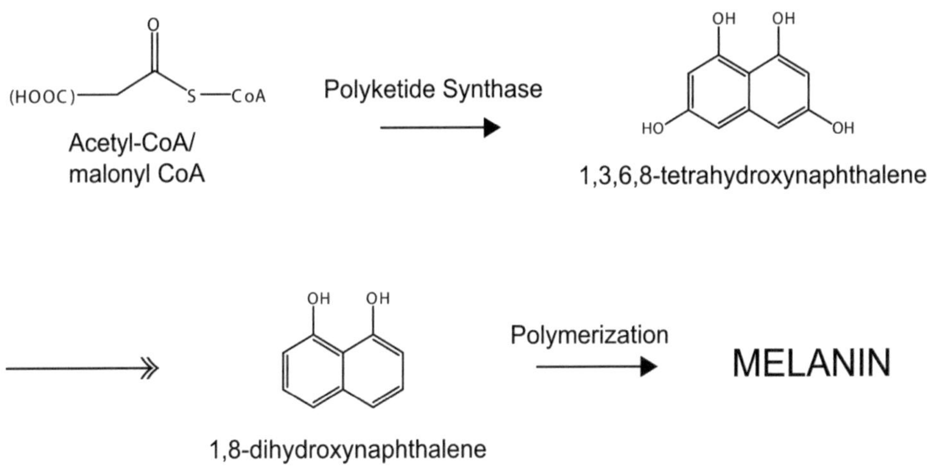

Figure 1. The biosynthetic pathway of fungal dihydroxynaphthalene(DHN)-melanin. Scheme adapted from Ref. [13].

However, some species of this class, including *Cladosporium resinae, Epicoccum nigrum, Hendersonula toruloidea, Eurotium echinulatum, Humicola grisea,* and *Hypoxylon archeri*, do not produce this type of pigment [11, 28, 86–88]. In the genus *Aspergillus*, DHN-melanin has not been identified in some members, as *A. nidulans* and *A. niger*. Bull [89] identified dopachrome (indole 5,6-quinone 2-carboxylic acid) and melanochrome (indole 5,6-quinone), which are intermediates in the DOPA-melanin pathway, in *A. nidulans* mutants defective in the production of melanin. Other studies confirmed the indolic nature of the melanin produced by *A. nidulans* [11, 90]. In *A. nidulans* strains, one tyrosinase was identified as the enzyme responsible for the production of melanin pigment, based on its substrate specificity (DOPA substrate) and susceptibility to inhibitors [91, 92]. In a recent study, our group characterized the pigment produced by *A. nidulans* mutants as DOPA-melanin according to the results obtained with specific inhibitors of DHN- and DOPA-melanin pathways [93].

The production of DOPA-melanin has also been investigated in other fungi such as *Neurospora crassa* [94], *Podospora anserina* [95], *A. nidulans* [91], *A. oryzae*, and *C. neoformans* [96]. A biosynthesis pathway for fungal DOPA-melanin, proposed by [11], is shown in **Figure 2**, which strongly resembles the pathway found in mammalian cells, though some of the details may differ.

In this pathway, there are two possible starting molecules, L-dopa and tyrosine. If L-dopa is the precursor molecule, it is oxidized to dopaquinone by laccase. If tyrosine is the precursor, it is first converted to L-dopa and then dopaquinone. The same enzyme, tyrosinase, carries out both steps. Dopaquinone, a highly reactive intermediate, forms leucodopachrome, which is then oxidized to dopachrome. Hydroxylation (and decarboxylation) yields dihydroxyindoles, which can polymerize spontaneously to form DOPA-melanin [10, 27, 97].

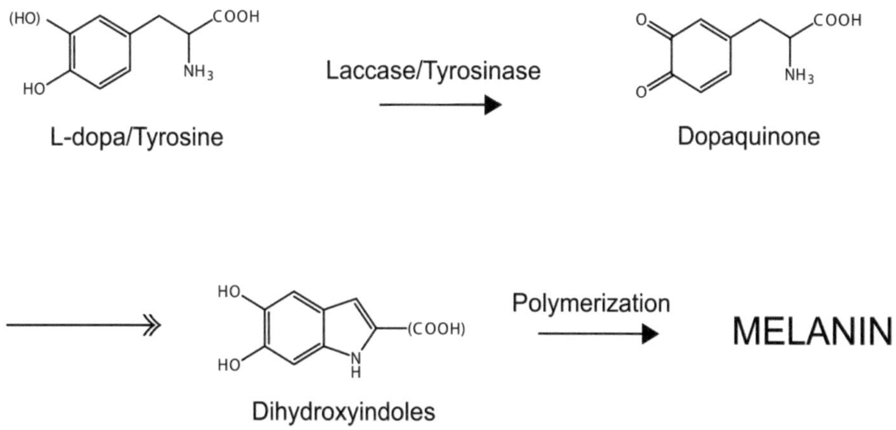

Figure 2. The biosynthetic pathway of the dihydroxyphenylalanine (DOPA)-melanin in fungi. Scheme adapted from Ref. [13].

Some fungi have more than one biosynthetic pathway of melanins. For example, *Aspergillus fumigatus* synthesizes DHN-melanin [98] and also produces a second type of melanin, piomelanins, from homogentisic acid by the tyrosine degradation pathway that protects the cell wall of hyphae from ROS, and gray-green DHN-melanins determine the structural integrity of the cell wall of conidia and their adhesive properties [99]. In *Agaricus bisporus*, melanins are formed from DOPA by tyrosinase and from γ-glutaminyl-4-hydroxybenzene by peroxidase and phenolase [100].

The extracellular fungal melanin, which is found in culture fluids usually in the form of granules, can be formed from some culture components, which are autoxidized or are oxidized by phenoloxidases released from the fungus during autolysis [10, 11, 27].

4. Biological activities of melanin

Despite the difference in their origins, melanin pigments have a number of common characteristics that allow them to fulfill their protective function. Several biological functions of melanins are closely associated to their chemical composition and structure. The presence of unpaired electrons in the melanin structure is responsible for various properties, including antioxidant, semiconductor, optical, electronic, and radio- and photoprotective [19].

The effect of melanin enhancing the survival of fungi under adverse conditions is mainly due to its function as an extracellular redox buffer, which can neutralize oxidants generated by the fungus in response to environmental stress [19]. It has been reported that melanin contributes for virulence of *C. neoformans*, protecting the pathogen against free radicals generated immunologically [29]. In *W. dermatitidis* and *A. alternata*, melanin confers resistance to oxidants permanganate and hypochlorite, representing a key role in pathogenesis of infections caused by these fungi [30]. Studies have shown that melanin of zoopathogenic and phytopathogenic fungi is essential for their parasitizing, due to its antioxidant properties [101].

Melanin pigment extracted from several fungal species has shown the ability to scavenge free radicals (reactive nitrogen and oxygen species), becoming a potential natural antioxidant. Melanins produced by *Exophiala pisciphila* and *Aspergillus bridgeri* ICTF-201 exhibited a significant DPPH (2,2-diphenyl-1-picrylhydrazyl) radical scavenging activity comparable with that of synthetic melanin, indicating its antioxidant potential [102, 103]. Melanin produced by *Schizophyllum commune* showed high free radical scavenging activity in a dose-dependent manner, when the melanin concentration was increased from 10 to 50 μg, the scavenging activity was also increased from 87% to 96%, similar to those obtained using ascorbic acid (standard compound used to measure free radical scavenging activity) [34]. Melanin pigment of *Fonsecaea pedrosoi* has antioxidant potential by reducing Fe(III) to Fe(II), ensuring the balance of its redox chemical microenvironment and minimizing the effect of oxidation of fundamental structures on fungal growth [104]. Similar results were also observed for melanin from *Ophiocordyceps sinensis*, which proved to be an effective DPPH radical scavenger and a strong ferrous iron chelator [105]. The chelating power of fungal melanin can be explained by various functional groups present in the structure of this pigment, which provide an array of multiple nonequivalent binding sites for metal ions [14, 22].

It has been reported that substances acting as antioxidants protect cells from ROS-mediated DNA damage, which can result in mutation and subsequent carcinogenesis. The excess free radicals may attack cellular constituents, as the cell membrane, nucleic acid, protein, enzymes, and other biomolecules, by peroxidation, resulting in the severe damage of cell functions and subsequent serious deleterious effects on the organism [106]. It has been reported that melanin protects melanocytes and keratinocytes from the induction of DNA strand broken by hydrogen peroxide, indicating that this pigment also has an important antioxidant role in the skin [107]. Studies in our laboratory showed that melanin extracted from hyperpigment-productive mutant (MEL1) of *A. nidulans* has the ability to scavenge the biological oxidants, as HOCl, and may be a promising material in cosmetic formulations to protect the skin against possible oxidative damage [31].

There is experimental evidence that fungal melanin may also act as an anti-aging drug, due to its action in reducing the generation of free radicals, clearing away the free radicals produced in excess, and enhancing the activities of antioxidant enzymes. Studies have shown that one of the major causes of aging is the surplus free radicals produced during the oxidative metabolism in the human body [108]. It was demonstrated that the melanin produced by fungus *Lachnum singerianum* YM296 significantly inhibited the formation of lipid peroxidation products and slowed down the aging process, elevating the levels of superoxide dismutase, glutathione peroxidase, and catalase and decreasing the level of malondialdehyde in mice liver and brain homogenate and serum, suggesting that this pigment could be used as a new anti-aging drug [109].

Researches have also shown that some fungal melanin exhibits immunomodulatory activity through the inhibition of pro-inflammatory cytokine production in T lymphocytes and monocytes, as well as fibroblasts and endothelial cells [12, 110, 111]. During an inflammatory response, cells of the innate and acquired immune systems release a variety of mediators, such as nitric oxide (NO), tumor necrosis factor-α (TNF-α), interleukins (IL), and the reactive nitrogen and oxygen species, which are implicated in the pathogenesis of a number of inflammatory diseases [112]. [113] reported that treatment of macrophages activated in vitro with melanin from the fungus *F. pedrosoi* inhibited the production of nitric oxide and Th1 cytokines. The study performed by [114] showed that the expression of inducible nitric oxide synthase gene decreased and lower levels of cytokines, such as IL-12 and TNF-α, were observed when activated macrophages were incubated with melanized cells of the *Fonsecaea monophora* fungus. Our studies demonstrated that melanin extracted from a highly melanized mutant (MEL1) of *A. nidulans* inhibited NO production in LPS-stimulated macrophages, with a maximum response of 82% inhibition, and also showed a dose-dependent inhibitory effect on TNF-α production, reaching an inhibition of 51.86% at a melanin concentration of 100 μg/mL. These results suggest that melanin from *A. nidulans* has potential as an anti-inflammatory agent and may be used in the future for development of new drugs with therapeutic utility [32].

Some studies have proposed that fungal melanin exhibits anti-radiation activity in vivo and in vivo and then could be explored as a probable radioprotector [16, 115]. Since melanin has a stable free radical population, it is thought that the radioprotective properties of this pigment result from a combination of physical shielding and quenching of cytotoxic free radicals generated by radiation [18]. [116] showed that *Lachnum* extracellular melanin (LEM404) had strong anti-ultraviolet radiation activity because the survival rates of *Escherichia coli*, *Staphylococcus aureus*, and *Saccharomyces cerevisiae* under UV radiation were significantly increased after in vitro addition of LEM404. Compared with the control groups, the antioxidant defense systems, such as superoxide dismutase and glutathione peroxidase activities, were improved significantly in mice of experiment groups, and the reactive oxygen species detected by malondialdehyde content were decreased significantly. These results confirmed that fungal melanin could be used as component of photoprotective creams mainly for its free radical scavenging rather than its light absorption properties. The probable mechanisms of radioprotection by melanin appear to be modulated in pro-survival pathways, immune system, and prevention of oxidative stress. It was reported that melanin isolated from the fungus *G. simplex* reduced the radiation-induced overproduction of pro-inflammatory cytokines (IL-6 and TNF-α), which might help in the recovery from radiation injury by preventing the aggravation of inflammation and oxidative stress [33]. This study confirmed the possible use of melanin-coated nanoparticles for protecting against radiotoxicity during radioimmunotherapy [117].

Recent studies have demonstrated that, in addition to the ability of transferring electrons arising under the action of radiation, melanin also possesses ionic conductivity due to its ability to transform any type of radiation energy not only into heat but also use it for the maintenance of redox processes in cells [118]. It was assumed that melanin pigments, participating in redox reactions, are able to perceive the energy of radiation (UV, visible light, and radiation) and convert it into useful reducing power for metabolic processes. This hypothesis is supported by the discovery of melanized fungi in soils contaminated by radioactive nuclides and areas around the damaged Chernobyl nuclear reactors, which not only survive high radiation levels but also have enhanced growth upon exposure [16, 19, 119, 120]. Owing to its semiconductor property, melanin becomes a promising material for organic bioelectronic devices like transistors, sensors, and batteries [121].

Fungal melanins also exhibit growth inhibitory effect against various microorganisms. The extracellular melanin isolated from *S. commune* showed significant antibacterial activity against *E. coli*, *Proteus* sp., *Klebsiella pneumonia*, and *Pseudomonas fluorescens* and antifungal activity against dermatophytic fungi, *Trichophyton simii*, and *T. rubrum* [34]. The *A. auricula* melanin displayed inhibitory activity on biofilm formation of the three bacterial strains, *E. coli* K-12, *Pseudomonas aeruginosa* PAO1, and *P. fluorescens* P-3, and there was a proportional reduction in biofilm biomass with the increase in pigment concentration. Confocal laser scanning microscopy (CLSM) analyses showed that the three strains formed thick and compact biofilms when grown in the absence of pigment, but the presence of *A. auricula* melanin resulted in thinner and looser cell aggregations on surfaces instead of normal biofilm architecture. This study suggested that *A. auricular* melanin inhibits quorum-sensing (QS)-regulated biofilm formation in all strains tested without interfering with their growth [122]. Silver nanoparticles incorporated *Yarrowia lipolytica* melanin exhibited antimicrobial activity against the pathogen *Salmonella paratyphi*, and they were also effective at disrupting biofilms on polystyrene as well as glass surfaces [123]. These nanoparticles displayed excellent antifungal properties toward an *Aspergillus* sp. isolated from a wall surface, suggesting the application of these nanoparticles as effective paint additives. The melanin-silver nanostructures with broad-spectrum antimicrobial activity against food pathogens also have potential applicability in food processing and food packaging industries [124].

The anti-cell proliferation effect of fungal melanin in tumoral cell lines has already been demonstrated. [34] reported that the extracellular melanin produced by the fungus *S. commune* was effective against human epidermoid larynx carcinoma cell line (HEP-2) in a concentration-dependent manner, indicating its potential application in cancer chemoprevention and chemotherapy.

The evaluation of the effect of fungal melanin on non-tumor cells is also interesting because it may serve as alternative to acute in vivo toxicity testing, avoiding the indiscriminate use of animals. The melanin produced by *A. bridgeri* was evaluated in vitro cytotoxicity assay using cell lines TE 355.Sk derived from normal human skin fibroblasts and HEK-293 derived from human embryonic kidney cells, and no cytotoxicity was observed against the two cell lines [103]. In our studies, the toxicity of the melanin from *A. nidulans* was also evaluated due to its potential practical application as antioxidant and anti-inflammatory agent. The results showed that the viability of mouse macrophages remained greater than 90% when these cells were treated with a high melanin concentration (100 μg/mL), indicating that this pigment has low cytotoxicity [32]. We also showed that the toxicity of *A. nidulans* melanin on mouse fibroblast McCoy cell line, after metabolic activation with hepatic S9 microsomal fraction, was much lesser (CI_{50} = 413.4 ± 3.1 μg/mL) than known cytotoxic agents such as cyclophosphamide (CI_{50} = 15 ± 1.2 μg/mL). In this study, we demonstrated that this melanin pigment did not induce gene mutations in different strains of *Salmonella typhimurium* used in the Ames assay. Based on these results, we suggest that the melanin produced by *A. nidulans* does not cause significant damage to the cellular components and might be used in the future for development of new therapeutic drugs [32].

5. Biotechnological applications of melanin

With the current knowledge about physical and chemical properties and the broad spectrum of biological activities, fungal melanins have attracted growing interest for their potential use in the fields of biomedicine, dermocosmetics, nanotechnology, and materials science.

5.1. Bioelectronic applications

In recent years, the electronics industry has been driven to develop materials and components that are cheaper and more environmentally friendly. As melanin has characteristics of functional materials and bioorganic, a growing number of researchers in the fields of materials science and organic electronics see the melanin with great interest, taking advantage of their properties for applications in organic electronic devices. Melanins present interesting optoelectronic properties, such as high optical absorption in the UV-Vis range, good transmission electronic, and ionic conductivity appreciably, pointing this biomaterial as a promising active component in organic electronic devices with low environmental impact [118, 121, 125–127].

Among the physical properties of melanin, the electrical conductivity is one of the most interesting to investigate in the perspective of technological application. The electrical conductivity properties of this biopolymer are similar to those of amorphous semiconductor solids, and then it can be considered an organic semiconductor, which is largely available and biocompatible and, consequently, cheaper and easier to process with respect to inorganic semiconductors, as silicon germanium. In particular, it can be considered a promising material for sensors and photovoltaic devices, due to broadband spectral absorbance and charge transport properties [128].

The technical literature describes the integration of organic semiconducting polymers as melanin in silicon electronic devices in view of the possibility of achieving multifunctional systems that combine electrical and optical properties of semiconductors, the structural versatility and mechanical characteristics of materials, and processing polymeric [129]. The production of devices based on thin film melanin exhibited electrical conductivity comparable to that of amorphous silicon [130]. In this study, melanin films showed excellent thermal stability and adhere well to glass substrates and silicon, indicating the possibility of using this technique for the production of films from synthetic melanin. Other groups have published various device architectures with applications such as memory (metal-insulator-semiconductor geometries) [131], batteries [132], and biomimetic interfaces [133].

Deposition of homogeneous melanin layers for optoelectronics application is an issue of considerable technological relevance. Synthetic melanin thin films deposited by spray-coating presented features ascribed to an amorphous semiconducting material [134]. They also showed that further improvement of conductivity together with an increased absorption in the NIR region, by doping the synthetic melanin macromolecule, could make this material a good candidate for optical sensing applications. It has been reported that the iron-melanin coating markedly enhances the catalytic activity of the gold nanoparticles (AuNPs) for both the hydrogen peroxide electroreduction and hydrogen evolution reaction [135]. This strategy may be used to improve nanomaterials with potential applications as efficient catalysts and electrocatalysts. Studies have shown that synthetic melanin-like nanoparticles complexed with paramagnetic Fe^{3+} ions have potential as a highly efficient and nontoxic contrast agent for magnetic resonance imaging instead of Gd^{3+}-based contrast agents, which can cause nephrotoxicity [136].

The optical and electronic properties of melanin have attracted the attention of researchers for the production of continuous thin films from conventional synthetic melanin, which have been used for a number of different device configurations, including chemi-sensors, next-generation solar cells, and a range of other detectors [126, 130, 134]. Potential also exists to use melanin films as an effective radiation sensitizer that could greatly improve the spectral range and efficiency of superconducting transition-edge bolometers [137].

The metal chelation properties of melanin offer interesting possibilities for melanin-based metal ion sensing. A piezoelectric sensor system capable of real-time detection of metal ions was constructed by cross-linking melanin onto the gold electrode of quartz crystal microbalance (QCM) and showed high sensitivity and selectivity to metal ions particularly for Hg(II) [138].

Melanin has many other interesting properties, such as ultraviolet absorption, which has been utilized to prepare optical lenses or filters. Studies have shown that it is possible to use melanin as an ultraviolet, visible and near-infrared absorbing pigment in opthalmic devices, protective eyewear, windows, packaging material, umbrellas, canopies, and other similar media suitable for providing protection from radiation [139, 140]. The incorporation of the melanin in solid plastic films of polyvinyl alcohol (PVA-melanin film system) to be used in conjunction with other plastics to make laminated sheets or lenses, including sunglasses, ski goggles, ophthalmic prescription lenses, helmets, windows, light filters for artificial lighting, and other light filters that protect people from potentially damaging UV and high-energy visible light has also been reported [141].

5.2. Medical applications

Despite its high biocompatibility, the use of melanin as a novel biomaterial in pharmaceutical and biomedical applications reported in literature is still scarce. A study performed with melanin nanoparticles as biocompatible drug nanocarriers, using metronidazole (antibiotic drug), showed that melanin could be a very interesting nanocarrier drug release device because it strongly responds to pH, being a very interesting feature for the treatment of intestine and colon diseases, which would greatly benefit with pH targeting [142]. Another study showed that systemic melanin-covered nanoparticle (MN) administration reduced hematologic toxicity in mice treated with radiation and that these structures provide efficient protection to bone marrow against radiotoxicity during radioimmunotherapy and in some cases external beam radiation therapy, permitting the administration to tumors of significantly higher doses [117].

Melanin has also been used to treat various types of malignant cancer tumors, disorders of the immune system including AIDS, diseases of blood origin and disorders due to the disturbances in cell homeostasis, and complex and hardly curable mental disorders (schizophrenia, epilepsy) involving nervous and other regulatory systems. A study on the use of melanin for the treatment of Parkinson's disease, an amelioration in the monkeys' overall functional ability and secondary motor manifestations by the administration of an effective amount of melanin in monkeys treated with MPTP (1-Methyl-4-phenyl-1,2,5,6-tetrahydropyridine), a toxin that causes a neurodegenerative disease, was observed. This study demonstrated that toxin-induced Parkinson's disease could be prevented in the melanin-treated animals because the administered melanin causes chelation or scavenging of toxins, such as MPTP, thus preventing a neurodegenerative disease, such as Parkinson's disease. The results of this study also showed that melanin administration to aid the recovery of neurons in a mammal having neuron injury suggests that melanin can be used to treat Alzheimer's disease [143].

Owing to their ability to increase the permeability of the blood-brain barrier, the melanin is also useful as carriers for other therapeutic agents, which must reach brain tissue to produce their therapeutic responses [144]. Two examples of such therapeutic agents that will cross the blood-brain barrier when linked to melanin are boron and nerve growth factor. According to the same authors, the melanin is also an effective vehicle for the transport of boron to cancerous sites in the body, mainly when the cancerous cells to be treated are located in the brain, because this pigment binds boron very strongly. The melanin can also function as a carrier for nerve growth factor due to the ability to get nerve growth factor across the blood-brain barrier, and this is the major advantage over conventional therapy.

In recent years, efforts have been focused on investigating the potential use of this pigment as active material in tissue repair engineering. Bettinger et al. [145] reported that thin films of melanin were found to enhance Schwann cell growth and neurite extension in rat pheochromocytoma cells (PC12 cells) compared to collagen films in vitro. Melanin films also induced an inflammation response that was comparable to silicone implants in vivo, and the implants were significantly resorbed after 8 weeks. These results showed that melanin thin films have great potential in the reconstruction of tissues, being biodegradable, and possess inflammatory response comparable to silicone. Another study of the biocompatibility of melanin thin films demonstrated that the melanin film effectively supports the growth of undifferentiated stem cells and their differentiation into neuronal precursors and neurons [146]. They related that high-quality melanin thin films display appealing features, such as reversible conductivity by controlled hydration—dehydration steps—excellent biocompatibility with stem cells, and water-resistant adhesion, for bioelectronic applications, e.g., in organic electrochemical transistors (OECTs), which can translate cellular activity into electrical signals [125, 147]. It has also been reported that melanin thin films possess highly desirable physical and biological properties that make them ideal for organic bioelectronic devices [130].

In cosmetic industry, there are great interests in the melanin, especially to protect against the noxious effects of UV radiation by incorporation in skin photoprotection formulations [35, 148]. The protective action of melanin is related to its high efficiency to absorb and scatter photons, particularly the higher-energy photons from the UV and blue part of the solar spectrum. Very likely, melanin photoprotection is also due to its ability to quench excited states of certain molecules and scavenge ROS that may be generated in pigmented cells [126]. Development of methods for producing melanin soluble in aqueous cosmetic buffers at physiological pH and temperature may make possible the use of this pigment as ingredients of face and hand creams, lotions, antiaging ointments, or foundation makeups, acting as a screen and antioxidant for the protection against photoinduced skin damages [149]. Other dermocosmetic applications of melanins include the use of the pigment for hair dyeing and the development of novel strategies for hair recoloration [150].

Since melanin has a stable free radical population, it is thought that the radioprotective properties of this pigment result from a combination of physical shielding and quenching of cytotoxic free radicals generated by radiation [18]. Some studies suggest the possible use of melanin-coated nanoparticles in medicine, mainly for protecting patients against the harmful effects of gamma rays during radioimmunotherapy [34, 151]. Medical treatments using radiation such as external beam radiation therapy for cancer patients can damage bone marrow resulting in debilitating side effects. In experimental models, melanin can successfully shield bone marrow from such side effects. Mice treated with melanin-coated nanoparticles have higher white blood cell and platelet counts than control mice after radiation treatment [117]. It has been reported the use of melanin, a biopolymer with good biocompatibility and biodegradability, intrinsic photo-acoustic properties, binding ability to drugs, and chelating property to radioactive metal ions, as an efficient endogenous nanosystem for imaging-guided chemotherapy [152]. According to the authors, melanin nanoparticles could successfully enter into the tumor and act as an efficient drug-delivery system, thereby greatly increasing the safe utility of the drugs for tumor treatment and significantly lowering the dosage used and its side effects.

A valuable biotechnological approach to the melanin-mediated synthesis of silver nanostructures with broad-spectrum antimicrobial activity has been developed. Silver nanostructures synthesized with melanin derived from *Y. lipolytica* displayed excellent antifungal activity against an *Aspergillus* sp. isolated from a wall surface, indicating its potential application as effective paint additives [123]. The melanin-mediated nanostructures with broad-spectrum antimicrobial activity against food pathogens may be considered suitable for many practical food packaging applications because they can effectively inhibit the growth of pathogens and increase the shelf life of packed food products [124].

5.3. Environmental applications

The chemical structure of melanin presents many oxygen-containing groups, including carboxyl, phenolic and alcoholic hydroxyl, carbonyl, and methoxy groups, which have the ability to bind to a broad spectrum of substances [153]. In literature, studies have confirmed that fungal melanin acts as metal chelators, enhancing the biomass-metal interaction and consequently its biosorption capacity [14]. Study conducted by [154] showed that a melanin-rich strain of the fungus *Cladosporium cladosporioides* biosorbed 2.5- to 4-fold more Ni, Cu, Zn, Cd, and Pb ion than non-melanic *Penicillium digitatum*. These authors also studied the culture of *C. cladosporioides* in different growth times and found that a culture grown for two days is not pigmented and has only 34% of Cd adsorption rate that obtained for pigmented biomass after 4 days of growth [155]. Another study reported that melanized fungus *Armillaria* adsorb high concentrations of cations from the surrounding environment; some ions (Al, Zn, Fe, Cu, and Pb) were 50–100 times more concentrated on rhizomorphs than in soil [156]. The results obtained in our laboratory using a melanin-overproducing mutant (MEL1) from *A. nidulans* fungus [31, 93] showed that biosorption capacity for neodymium and lanthanum varied with stage of growth of this mutant; the biomass obtained after 72 hours of growth exhibited a 75% increase compared to the biomass of 48 hours. This result is related to melanin production during growth of the MEL1 mutant, since the biomass 48 hours is slightly pigmented, while the 72 hours biomass is dark due to the increased production of pigment [157]. Therefore, the pigmented biomass of the MEL1 mutant may be considered as a promising biosorbents for removal/recovery of the rare earth elements from wastewater due to the presence of the melanin increase significantly metal complexing capacity, improving the efficiency of biosorption process [157].

Some melanized fungi have shown to be good candidates for bioremediation of contaminated sites, due to the ability of fungal melanin to bind to heavy metals and radionuclides in contaminated sites. Experimental evidence shows that the accumulation of ^{90}Sr by conidia or mycelium by a range of microfungal species is greater in pigmented than in unpigmented species [158]. [159] In a study on the uptake efficiency of the radiocesium (^{137}Cs) and radiocobalt (^{60}Co) in melanized and nonmelanized fungi, it was observed that 60% of both radionuclides were uptaken by melanin of *A. alternata* and *Aspergillus pulverulents*. These results can be explained by melanin or other natural pigments present in the cell wall of these fungi that can act as the radiation receptor and/or as an energy transporter for metabolism. Other studies have demonstrated the potential application of the melanized fungi for the removal of radionuclides and heavy metals from aqueous solutions, providing an alternative means to affect cleanup of industrial effluent [16, 120, 160–164]. It has been reported that fungal melanin arranged in nanoparticles protects against extremely high levels of ionizing radiation and suggests that the protective efficacy of this pigment is a function of its chemical structure, the presence of stable free radical, and spatial arrangement [18]. According to the authors, these nanoshells have the potential use for environmental bioremediation, for example, to prevent the spread of radioactive contamination to ground water because the melanin is expected to encapsulate the radioactive particles and thereby reduce their spread. In this way, melanin nanoshells may be used to contain radiation from radioactive waste and biomedical radioactive materials.

6. Conclusion

Melanin possesses physicochemical properties and biological activities that make it a suitable biomaterial for a wide range of applications in cosmetic, pharmaceutical, electronic, and food processing industries. In addition, this pigment has a considerable interest biotechnological because it can be produced on a large scale with low cost, making its use for future practical applications economically advantageous. However, it is necessary to expand the knowledge about the structure-property-function relationships for the development of melanin-based technology. In the context, we hope that the information in this book will be useful and will encourage a greater number of researches on fungal melanin, which might be useful to deploy innovative and sustainable solutions for human health and the environment.

Author details

Sandra R. Pombeiro-Sponchiado[1]*, Gabriela S. Sousa[1], Jazmina C. R. Andrade[1], Helen F. Lisboa[2] and Rita C. R. Gonçalves[3]

*Address all correspondence to: sponchia@iq.unesp.br

1 Department of Biochemistry and Technology Chemistry, Institute of Chemistry, UNESP, São Paulo State University, Araraquara, São Paulo, Brazil

2 Institute of Naturals and Exacts Sciences, UFMT, Mato Grosso Federal University, Rondonopolis, Mato Grosso, Brazil

3 Department of Pharmaceutical Sciences, UFES, Espirito Santo Federal University, Vitória, Espirito Santo, Brazil

References

[1] Lopes FC, Tichota DM, Pereira JQ, Segalin J, De Oliveira Rios A, Brandelli A. Pigment production by filamentous fungi on agro-industrial byproducts: An eco-friendly alternative. Appl Biochem Biotechnol. 2013;171(3):616–625.

[2] Demain AL. Importance of microbial natural products and the need to revitalize their discovery. J Ind Microbiol Biotechnol. 2014;41(2):185–201.

[3] Kirti K, Amita S, Priti S, Kumar AM, Jyoti S. Colorful world of microbes: Carotenoids and their applications. Adv Biol. 2014;2014(1):1–13.

[4] Kumar A, Vishwakarma HS, Singh J, Kumar M. Microbial pigments: Production and their applications in various industries. Int J Pharm Chem Biol Sci. 2015;5(1):203–212.

[5] Panesar R, Kaur S, Panesar PS. Production of microbial pigments utilizing agro-industrial waste: A review. Curr Opin Food Sci 2015;1:70–76. Available from: http://dx.doi.org/10.1016/j.cofs.2014.12.002

[6] Akilandeswari P, Pradeep BV. Exploration of industrially important pigments from soil fungi. Appl Microbiol Biotechnol. 2016;100(4):1631–1643.

[7] Duran N, Teixeira MF, De Conti R, Esposito E Ecological-friendly pigments fungi. Critic Rev Food Sci Nutr. 2002;42(1):53–66.

[8] Dufossé L, Fouillaud M, Caro Y, Mapari SAS, Sutthiwong N. Filamentous fungi are large-scale producers of pigments and colorants for the food industry. Curr Opin Biotechnol. 2014;26:56–61.

[9] Chambergo FS, Valencia EY. Fungal biodiversity to biotechnology. Appl Microbiol Biotechnol. 2016;100(6):2567–2577.

[10] Butler MJ, Day AW. Fungal melanins: A review. Can J Microbiol. 1998;44(12):1115–1136.

[11] Bell AA, Wheeler MH. Biosynthesis and functions of fungal melanins. Annu Rev Phytopathol. 1986;24:411–451.

[12] Plonka PM, Grabacka M. Melanin synthesis in microorganisms – biotechnological and medical aspects. Acta Biochim Pol. 2006;53(3):429–443.

[13] Eisenman HC, Casadevall A. Synthesis and assembly of fungal melanin. Appl Microbiol Biotechnol. 2012;93(3):931–940.

[14] Fogarty RV, Tobin JM. Fungal melanins and their interaction with metals. Enzyme Microb Technol. 1996;19(June 1995):317.

[15] García-Rivera J, Casadevall A. Melanization of *Cryptococcus neoformans* reduces its susceptibility to the antimicrobial effects of silver nitrate. Med Mycol. 2001;39(October 2000):353–357.

[16] Dadachova E, Bryan RA, Huang X, Moadel T, Schweitzer AD, Aisen P, et al.. Ionizing radiation changes the electronic properties of melanin and enhances the growth of melanized fungi. PLoS One. 2007;2(5):1–13.

[17] Gómez BL, Nosanchuk JD. Melanin and fungi. Curr Opin Infect Dis. 2003;16(2):91–96.

[18] Dadachova E, Bryan RA, Howell RC, Schweitzer AD, Aisen P, Nosanchuk JD, et al.. The radioprotective properties of fungal melanin are a function of its chemical composition, stable radical presence and spatial arrangement. Pigment Cell Melanoma Res. 2008;21(2):192–199.

[19] Gessler NN, Egorova AS, Belozerskaya TA. Melanin pigments of fungi under extreme environmental conditions (Review). Appl Biochem Microbiol. 2014;50(2):105–113.

[20] Schnitzler N, Peltroche-Llacsahuanga H, Bestier N, Zündorf J, Lütticken R, Haase G. Effect of melanin and carotenoids of *Exophiala (Wangiella) dermatitidis* on phagocytosis, oxidative burst, and killing by human neutrophils. Infect Immun. 1999;67(1):94–101.

[21] Romero-Martinez R, Wheeler M, Guerrero-Plata A, Rico G, Torres-Guerrero H. Biosynthesis and functions of melanin in *Sporothrix schenckii*. Infect Immun. 2000;68(6):3696–3703.

[22] Nosanchuk JD, Stark RE, Casadevall A. Fungal melanin: What do we know about structure? Front Microbiol. 2015;6:1–7.

[23] Doering TL, Nosanchuk JD, Roberts WK, Casadevall A. Melanin as a potential cryptococcal defence against microbicidal proteins. Med Mycol. 1999;37(3):175–181.

[24] Casadevall A, Rosas AL, Nosanchuk JD. Melanin and virulence in *Cryptococcus neoformans*. Curr Opin Microbiol. 2000;3(4):354–358.

[25] Rosas ÁL, Casadevall A. Melanization decreases the susceptibility of *Cryptococcus neoformans* to enzymatic degradation. Mycopathologia. 2001;151(2):53–56.

[26] Paolo WF, Dadachova E, Mandal P, Casadevall A, Szaniszlo PJ, Nosanchuk JD. Effects of disrupting the polyketide synthase gene WdPKS1 in *Wangiella [Exophiala] dermatitidis* on melanin production and resistance to killing by antifungal compounds, enzymatic degradation, and extremes in temperature. BMC Microbiol. 2006;6:55.

[27] Langfelder K, Streibel M, Jahn B, Haase G, Brakhage AA. Biosynthesis of fungal melanins and their importance for human pathogenic fungi. Fungal Genet Biol. 2003;38(2):143–158.

[28] Nosanchuk JD, Casadevall A. The contribution of melanin to microbial pathogenesis. Cell Microbiol. 2003;5(4):203–223.

[29] Wang Y, Casadevall A. Susceptibility of melanized and nonmelanized *Cryptococcus neoformans* to nitrogen- and oxygen-derived oxidants. Infect Immun. 1994;62(7):3004–3007.

[30] Jacobson ES, Hove E, Emery HS. Antioxidant function of melanin in black fungi. Infect Immun. 1995;63(12):4944–4945.

[31] Goncalves CRR, Pombeiro-Sponchiado SR. Antioxidant activity of the melanin pigment extracted from *Aspergillus nidulans*. Biol Pharm Bull. 2005;28(6):1129–1131.

[32] Gonçalves R de CR, Kitagawa RR, Raddi MSG, Carlos IZ, Pombeiro-Sponchiado SR. Inhibition of nitric oxide and tumour necrosis factor-α production in peritoneal macrophages by *Aspergillus nidulans* melanin. Biol Pharm Bull. 2013;36(12):1915–1920.

[33] Kunwar A, Adhikary B, Jayakumar S, Barik A, Chattopadhyay S, Raghukumar S, et al.. Melanin, a promising radioprotector: Mechanisms of actions in a mice model. Toxicol Appl Pharmacol. 2012;264(2):202–211.

[34] Arun G, Eyini M, Gunasekaran P. Characterization and biological activities of extracellular melanin produced by *Schizophyllum commune* (Fries). Indian J Exp Biol. 2015;53(6):380–387.

[35] d'Ischia M, Wakamatsu K, Cicoira F, Di Mauro E, Garcia-Borron JC, Commo S, et al.. Melanins and melanogenesis: From pigment cells to human health and technological applications. Pigment Cell Melanoma Res. 2015;28(5):520–544.

[36] Demain AL, Adrio JL. Strain improvement for production of pharmaceuticals and other microbial metabolites by fermentation. Nat Compd Drugs. 2008;1:251–289.

[37] Babitha S. Microbial Pigments. In: Nigam PSN, Pandey A, editors. Biotechnology for agro-industrial residues utilization: utilization of agro-residues. Netherlands: Springer; 2009. p. 147–162.

[38] Meyer V, Wu B, Ram AFJ. Aspergillus as a multi-purpose cell factory: Current status and perspectives. Biotechnol Lett. 2011;33(3):469–476.

[39] Mapari SAS, Nielsen KF, Larsen TO, Frisvad JC, Meyer AS, Thrane U. Exploring fungal biodiversity for the production of water-soluble pigments as potential natural food colorants. Curr Opin Biotechnol. 2005;16(2):231–238.

[40] Mapari SA, Meyer AS, Thrane U, Frisvad JC. Identification of potentially safe promising fungal cell factories for the production of polyketide natural food colorants using chemotaxonomic rationale. Microb Cell Fact. 2009;8:24.

[41] Dikshit R, Tallapragada P. *Monascus purpureus*: A potential source for natural pigment production. J Microbiol Biotechnol Res. 2011;1(4):164–174.

[42] Marova I, Carnecka M, Halienova A, Certik M, Dvorakova T, Haronikova A. Use of several waste substrates for carotenoid-rich yeast biomass production. J Environ Manage. 2012;95(Suppl.):S338–S342.

[43] Sharmila G, Nidhi B, Muthukumaran C. Sequential statistical optimization of red pigment production by *Monascus purpureus* (MTCC 369) using potato powder. Ind Crops Prod. 2013;44:158–164.

[44] Prajapati VS, Soni N, Trivedi UB, Patel KC. An enhancement of red pigment production by submerged culture of *Monascus purpureus* MTCC 410 employing statistical methodology. Biocatal Agric Biotechnol. 2014;3(2):140–145.

[45] Hajjaj H, Goma G, François JM. Effect of the cultivation mode on red pigments production from *Monascus ruber*. Int J Food Sci Technol. 2015;50(8):1–6.

[46] Zhang M, Xiao G, Thring RW, Chen W, Zhou H, Yang H. Production and characterization of melanin by submerged culture of culinary and medicinal fungi *Auricularia auricula*. Appl Biochem Biotechnol. 2015;176(1):253–266.

[47] da Costa Souza PN, Grigoletto TLB, de Moraes LAB, Abreu LM, Guimarães LHS, Santos CR, et al.. Production and chemical characterization of pigments in filamentous fungi. Microbiology. 2015;162:12–22.

[48] Joshi VK, Attri D, BaJa A, Bhushan S. Microbial pigments. Indian J Biotechnol. 2003;2(Aug):362–369.

[49] Ahn J, Jung J, Hyung W, Haam S, Shin C. Enhancement of *Monascus* pigment production by the culture of *Monascus* sp. J101 at low temperature. Biotechnol Prog. 2006;22(1):338–340.

[50] Lisboa HCF.Influence of culture conditions on the production of melanin pigment by *Aspergillus fungus* [thesis]. Araraquara: Sao Paulo State University; 2003.

[51] Orozco SFB, Kilikian BV. Effect of pH on citrinin and red pigments production by *Monascus purpureus* CCT3802. World J Microbiol Biotechnol. 2008;24(2):263–268.

[52] Kang B, Zhang X, Wu Z, Wang Z, Park S. Production of citrinin-free *Monascus* pigments by submerged culture at low pH. Enzyme Microb Technol. 2014;55:50–57.

[53] Tudor D, Robinson SC, Cooper PA. The influence of pH on pigment formation by lignicolous fungi. Int Biodeterior Biodegrad. 2013;80:22–28.

[54] Méndez A, Pérez C, Montañéz JC, Martínez G, Aguilar CN. Red pigment production by *Penicillium purpurogenum* GH2 is influenced by pH and temperature. J Zhejiang Univ Sci B. 2011;12(12):961–968.

[55] Velmurugan P, Lee YH, Venil CK, Lakshmanaperumalsamy P, Chae JC, Oh BT. Effect of light on growth, intracellular and extracellular pigment production by five pigment-producing filamentous fungi in synthetic medium. J Biosci Bioeng. 2010;109(4):346–350.

[56] Zhou Z, Yin Z, Hu X. Corncob hydrolysate, an efficient substrate for *Monascus* pigment production through submerged fermentation. Biotechnol Appl Biochem. 2014;61(6):716–723.

[57] Said FM, Chisti Y, Brooks J, The effects of forced aeration and initial moisture level on red pigment and biomass production by *Monascus ruber* in packed bed solid state fermentation. Int J Environ Sci Dev. 2010;1(1):1–4.

[58] Ruiz B, Chávez A, Forero A, García-Huante Y, Romero A, Sánchez M, et al.. Production of microbial secondary metabolites: Regulation by the carbon source. Crit Rev Microbiol. 2010;36(2):146–167.

[59] Pradeep FS, Begam MS, Palaniswamy M, Pradeep BV. Influence of culture media on growth and pigment production by *Fusarium moniliforme* KUMBF1201 isolated from paddy field soil. World Appl Sci J. 2013;22(1):70–77.

[60] Gunasekaran S, Poorniammal R. Optimization of fermentation conditions for red pigment production from *Penicillium* sp. under submerged cultivation. Afr J Biotechnol. 2008;7(12):1894–1898.

[61] nee'Nigam PS. Production of bioactive secondary metabolites. In: Nigam PSN, Pandey A, editors. Biotechnology for agro-industrial residues utilization: utilization of agro-residues. Netherlands: Springer; 2009. p. 129-145.

[62] Quereshi S, Pandey AK, Singh J. Optimization of fermentation conditions for red pigment production from *Phoma herbarum* (FGCC# 54) under submerged cultivation. J Phytol. 2010;2(9).

[63] Celestino JR, de Carvalho LE, da Paz Lima, M, Lima AM, Ogusku MM, de Souza JVB. Bioprospecting of Amazon soil fungi with the potential for pigment production. Process Biochem. 2014;49(4):569–575.

[64] Jalmi P, Bodke P, Wahidullah S, Raghukumar S. The fungus *Gliocephalotrichum simplex* as a source of abundant, extracellular melanin for biotechnological applications. World J Microbiol Biotechnol. 2012;28(2):505–512.

[65] Sun S, Zhang X, Chen W, Zhang L, Zhu H. Production of natural edible melanin by *Auricularia auricula* and its physicochemical properties. Food Chem. 2016;196:486–492.

[66] Pandey A, Soccol CR, Mitchell D. New developments in solid state fermentation: I-bioprocesses and products. Process Biochem. 2000;35(10):1153–1169.

[67] Singhania RR, Soccol CR, Pandey A. Application of tropical agro-industrial residues as substrate for solid-state fermentation processes. In: Pandey A, Soccol CR, Larroche C, editors. Current developments in solid-state fermentation. New York: Springer; 2008. p. 412–442.

[68] Velmurugan P, Hur H, Balachandar V, Kamala-Kannan S, Lee KJ, Lee SM, Chae JC, Oh BT. *Monascus* pigment production by solid-state fermentation with corn cob substrate. J Biosci Bioeng. 2011;112(6):590–594.

[69] Babitha S, Soccol CR, Pandey A Jackfruit seed-a novel substrate for the production of *Monascus* pigments through solid-state fermentation. Food Technol Biotechnol. 2006;44(4):465–471.

[70] Hamano PS, Kilikian BV. Production of red pigments by *Monascus ruber* in culture media containing corn steep liquor. Brazilian J Chem Eng. 2006;23(4):443–449.

[71] Silveira ST, Daroit DJ, Brandelli A. Pigment production by *Monascus purpureus* in grape Waste using factorial design. LWT – Food Sci Technol. 2008;41(1):170–174.

[72] Rani MHS, Ramesh T, Subramanian J, Kalaiselvam M. Production and characterization of melanin pigment from halophilic black yeast *Hortaea werneckii*. Int J Pharma Res Rev. 2013;2(8):9–17.

[73] Zou Y, Tian M. Fermentative production of melanin by *Auricularia auricula*. J Food Process Preserv. 2016;0:0–5.

[74] Pretti TS. Optimization of the culture conditions of *Aspergillus nidulans* fungus for the production of melanin using agroindustrial residues [thesis]. Araraquara: Sao Paulo State University, 2009.

[75] Enochs WS, Nilges MJ, Swartz HM. A standardized test for the identification and characterization of melanins using electron paramagnetic resonance (EPR) spectroscopy. Pigment Cell Res. 1993;6(2):91–99.

[76] Crippa R, Horak V, Prota G, Svoronos P, Wolfram L. Chemistry of melanins alkaloids. Chem Pharmacol. 1990;36:253–323.

[77] Wilczok T, Bilińska B Buszman E, Kopera M. Spectroscopic studies of chemically modified synthetic melanins. Arch Biochem Biophys. 1984;231(2):257–262.

[78] Duff GA, Roberts JE, Foster N. Analysis of the structure of synthetic and natural melanins by solid-phase NMR. Biochemistry. 1988;27(18):7112–7116.

[79] Piattelli M, Fattorusso E, Nicolaus RA, Magno S. The structure of melanins and melanogenesis—V: Ustilagomelanin. Tetrahedron. 1965;21(11):3229–3236.

[80] Stüssi H, Rast DM. The biosynthesis and possible function of γ-glutaminyl-4-hydroxybenzene in *Agaricus bisporus*. Phytochemistry. 1981;20(10):2347–2352.

[81] Eisenman HC, Mues M, Weber SE, Frases S, Chaskes S, Gerfen G, Casadevall A. *Cryptococcus neoformans* laccase catalyses melanin synthesis from both D-and L-DOPA. Microbiology. 2007;153(12):3954–3962.

[82] Frases S, Salazar A, Dadachova E, Casadevall A *Cryptococcus neoformans* can utilize the bacterial melanin precursor homogentisic acid for fungal melanogenesis. Appl Environ Microbiol. 2007;73(2):615–621.

[83] Garcia-Rivera J, Eisenman HC, Nosanchuk JD, Aisen P, Zaragoza O, Moadel T, et al.. Comparative analysis of *Cryptococcus neoformans* acid-resistant particles generated from pigmented cells grown in different laccase substrates. Fungal Genet Biol. 2005;42(12):989–998.

[84] Geis PA, Wheeler MH, Szaniszlo PJ. Pentaketide metabolites of melanin synthesis in the dematiaceous fungus *Wangiella dermatitidis*. Arch Microbiol. 1984;137(4):324–328.

[85] Wheeler MH, Stipanovic RD. Melanin biosynthesis and the metabolism of flaviolin and 2-hydroxyjuglone in *Wangiella dermatitidis*. Arch Microbiol. 1985;142(3):234–241.

[86] Saiz-Jimenez C. Microbial melanins in stone monuments. Sci Total Environ. 1995;167(1):273–286.

[87] Henson JM, Butler MJ, Day AW. The dark side of the mycelium: Melanins of phytopathogenic fungi. Annu Rev Phytopathol. 1999;37(1):447–471.

[88] Wu Y, Shan L, Yang S, Ma A. Identification and antioxidant activity of melanin isolated from *Hypoxylon archeri*, a companion fungus of *Tremella fuciformis*. J Basic Microbiol. 2008;48(3):217–221.

[89] Bull AT. Chemical composition of wild-type and mutant *Aspergillus nidulans* cell walls. The nature of polysaccharide and melanin constituents. J Gen Microbiol. 1970;63(1):75–94.

[90] Rowley BI, Pirt SJ. Melanin production by *Aspergillus nidulans* in batch and chemostat cultures. Microbiology. 1972;72(3):553–563.

[91] Bull AT, Carter BLA. The isolation of tyrosinase from *Aspergillus nidulans*, its kinetic and molecular properties and some consideration of its activity in vivo. Microbiology. 1973;75(1):61–73.

[92] Martinelli SD, Bainbridge BW. Phenoloxidases of *Aspergillus nidulans*. Trans Br Mycol Soc. 1974;63(2):361–370.

[93] Gonçalves RCR, Lisboa HCF, Pombeiro-Sponchiado SR. Characterization of melanin pigment produced by *Aspergillus nidulans*. World J Microbiol Biotechnol. 2012;28(4):1467–1474.

[94] Fling M, Horowitz NH, Heinemann SF. The isolation and properties of crystalline tyrosinase from *Neurospora*. J Biol Chem. 1963;238(6):2045–2053.

[95] Esser K. Phenoloxidases in the ascomycete *Podospora anserina*. I. The identification of laccase and tyrosinase in the wild strain. Arch Mikrobiol. 1963;46:217–226.

[96] Williamson PR, Wakamatsu K, Ito S. Melanin biosynthesis in *Cryptococcus neoformans*. J Bacteriol. 1998;180(6):1570–1572.

[97] Prota G. Melanins and Melanogenesis. Academic Press, San Diego, CA; 1992.

[98] Tsai HF, Fujii I, Watanabe A, Wheeler MH, Chang YC, Yasuoka Y, Ebizuka Y, Kwon-Chung KJ. Pentaketide melanin biosynthesis in *Aspergillus fumigatus* requires chain-length shortening of a heptaketide precursor. J Biol Chem. 2001;276(31):29292–29298.

[99] Schmaler-Ripcke J, Sugareva V, Gebhardt P, Winkler R, Kniemeyer O, Heinekamp T, Brakhage AA. Production of pyomelanin, a second type of melanin, via the tyrosine degradation pathway in *Aspergillus fumigatus*. Appl Environ Microbiol. 2009;75(2):493–503.

[100] Jolivet S, Arpin N, Wichers HJ, Pellon G. *Agaricus bisporus* browning: A review. Mycol Res. 1998;102(12):1459–1483.

[101] Gessler NN, Aver'yanov AA, Belozerskaya TA. Reactive oxygen species in regulation of fungal development. Biochemistry. 2007;72(10):1091–1109.

[102] Zhan F, He Y, Zu Y, Li T, Zhao Z. Characterization of melanin isolated from a dark septate endophyte (DSE), *Exophiala pisciphila,* World J Microbiol Biotechnol. 2011;27(10):2483–2489.

[103] Kumar CG, Mongolla P, Pombala S, Kamle A, Joseph J. Physicochemical characterization and antioxidant activity of melanin from a novel strain of *Aspergillus bridgeri* ICTF-201. Lett Appl Microbiol. 2011;53(3):350–358.

[104] Cunha MML, Franzen AJ, Seabra SH, Herbst MH, Vugman NV, Borba LP, Souza W, Rozental S. Melanin in Fonsecaea pedrosoi: A trap for oxidative radicals. BMC Microbiol. 2010;10(1):1–9.

[105] Dong C, Yao Y. Isolation, characterization of melanin derived from *Ophiocordyceps sinensis,* an entomogenous fungus endemic to the Tibetan Plateau. J Biosci Bioeng. 2012;113(4):474–479.

[106] Lopaczynski W, Zeisel SH. Antioxidants, programmed cell death, and cancer. Nutr Res. 2001;21(1):295–307.

[107] Hoogduijn MJ, Cemeli E, Anderson D, Wood JM, Thody AJ. Melanin protects against Ho-induced DNA strand breaks through its ability to bind Ca. Br J Dermatol. 2003;148(4):867.

[108] Harman D. Free radical theory of aging: An update. Ann N Y Acad Sci. 2006;1067(1):10–21.

[109] Lu Y, Ye M, Song S, Li L, Shaikh F, Li J. Isolation, purification, and anti-aging activity of melanin from *Lachnum singerianum*. Appl Biochem Biotechnol. 2014;174(2):762–771.

[110] Mohagheghpour N, Waleh N, Garger SJ, Dousman L, Grill LK, Tusé D. Synthetic melanin suppresses production of proinflammatory cytokines. Cell Immunol. 2000;199(1):25–36.

[111] Mednick AJ, Nosanchuk JD, Casadevall A. Melanization of *Cryptococcus neoformans* affects lung inflammatory responses during cryptococcal infection. Infect Immun. 2005;73(4):2012–2019.

[112] Cruvinel WDM, Mesquita Júnior D, Araújo JAP, Catelan TTT, Souza AWSD, Silva NPD, Andrade LEC. Immune system: Part I. Fundamentals of innate immunity with emphasis on molecular and cellular mechanisms of inflammatory response. Revista brasileira de reumatologia. 2010; 50(4):434–447.

[113] Bocca AL, Brito PP, Figueiredo F, Tosta CE. Inhibition of nitric oxide production by macrophages in chromoblastomycosis: A role for *Fonsecaea pedrosoi* melanin. Mycopathologia. 2006;161(4):195–203.

[114] Zhang J, Wang L, Xi L, Huang H, Hu Y, Li X, Huang X, Lu S, Sun J. Melanin in a meristematic mutant of Fonsecaea monophora inhibits the production of nitric oxide and Th1 cytokines of murine macrophages. Mycopathologia. 2013;175(5–6):515–522.

[115] Schweitzer AD, Howell RC, Jiang Z, Bryan RA, Gerfen G, Chen CC, Mah D, Cahill S, Casadevall A, Dadachova E. Physico-chemical evaluation of rationally designed melanins as novel nature-inspired radioprotectors. PLoS One. 2009;4(9):1–8.

[116] Ye M, Guo G, Lu Y, Song S, Wang HY, Yang L. Purification, structure and anti-radiation activity of melanin from *Lachnum* YM404. Int J Biol Macromol. 2014;63(1):170–176.

[117] Schweitzer AD, Revskaya E, Chu P, Pazo V, Friedman M, Nosanchuk JD, Cahill S, Frases S, Casadevall A, Dadachova E. Melanin-covered nanoparticles for protection of bone marrow during radiation therapy of cancer. Int J Radiat Oncol Biol Phys. 2010;78(5):1494–1502.

[118] Mostert AB, Powell BJ, Pratt FL, Hanson GR, Sarna T, Gentle IR, Meredith P. Role of semiconductivity and ion transport in the electrical conduction of melanin. Proc Natl Acad Sci. 2012;109(23):8943–8947.

[119] Mironenko NV, Alekhina IA, Zhdanova NN, Bulat SA. Intraspecific variation in gamma-radiation resistance and genomic structure in the filamentous fungus *Alternaria alternata*: A case study of strains inhabiting Chernobyl Reactor No. 4. Ecotoxicol Environ Saf. 2000;45(2):177–187.

[120] Dighton J, Tugay T, Zhdanova N. Fungi and ionizing radiation from radionuclides. FEMS Microbiol Lett. 2008;281(2):109–120.

[121] Mostert AB, Powell BJ, Gentle IR, Meredith P. On the origin of electrical conductivity in the bio-electronic material melanin. Appl Phys Lett. 2012;100(9).

[122] Bin L, Wei L, Xiaohong C, Mei J, Mingsheng D. In vitro antibiofilm activity of the melanin from *Auricularia auricula*, an edible jelly mushroom. Ann Microbiol. 2012;62(4):1523–1530.

[123] Apte M, Girme G, Bankar A, Ravikumar A, Zinjarde S. 3,4-dihydroxy-L-phenylalanine-derived melanin from *Yarrowia lipolytica* mediates the synthesis of silver and gold nanostructures. J Nanobiotechnol. 2013;3–11.

[124] Kiran GS, Dhasayan A, Lipton AN, Selvin J, Arasu MV, Al-Dhabi NA. Melanin-templated rapid synthesis of silver nanostructures. J Nanobiotechnol. 2014;12(1):18.

[125] Goncalves PJ, Baffa O, Graeff CFO, Gonçalves PJ, Filho OB. Effects of hydrogen on the electronic properties of synthetic melanin. J Appl Phys. 2006;99(10):104701.

[126] Meredith P, Sarna T. The physical and chemical properties of eumelanin. Pigment Cell Res. 2006;19(6):572–594.

[127] Ambrico M, Ambrico PF, Ligonzo T, Cardone A, Cicco SR, d'Ischia M, Farinola GM. From commercial tyrosine polymers to a tailored polydopamine platform: Concepts, issues and challenges en route to melanin-based bioelectronics. J Mater Chem C. 2015;3(25):6413–6423.

[128] Ligonzo T, Ambrico M, Augelli V, Perna G, Schiavulli L, Tamma MA, et al.. Electrical and optical properties of natural and synthetic melanin biopolymer. J Non Cryst Solids. 2009;355(22–23):1221–1226.

[129] Ambrico M, Ambrico PF, Cardone A, Ligonzo T, Cicco SR, Di Mundo R, et al.. Melanin layer on silicon: An attractive structure for a possible exploitation in bio-polymer based metal-insulator-silicon devices. Adv Mater. 2011;23(29):3332–3336.

[130] Bothma JP, de Boor J, Divakar U, Schwenn PE, Meredith P. Device-quality electrically conducting melanin thin films. Adv Mater. 2008;20(18):3539–3542.

[131] Ambrico M, Ambrico PF, Ligonzo T, Cardone A, Cicco SR, Lavizzera A, et al.. Memory-like behavior as a feature of electrical signal transmission in melanin-like bio-polymers. Appl Phys Lett. 2012;100(25).

[132] Kim YJ, Wu W, Chun SE, Whitacre JF, Bettinger CJ. Biologically derived melanin electrodes in aqueous sodium-ion energy storage devices. Proc Natl Acad Sci. 2013;110(52):20912–20917.

[133] Ambrico M, Ambrico PF, Cardone A, Cicco SR, Palumbo F, Ligonzo T, Mundo RD, Petta V, Augelli V, Favia P, Farinola GM. Melanin-like polymer layered on a nanotextured silicon surface for a hybrid biomimetic interface. J Mater Chem C. 2014;2(3):573–582.

[134] Morresi L, Ficcadenti M, Pinto N, Murri R, Cuccioloni M, Angeletti M, et al.. Optical and electrical behavior of synthetic melanin thin films spray-coated. Energy Procedia. 2010;2(1):177–182.

[135] Orive AG, Grumelli D, Vericat C, Ramallo-López JM, Giovanetti L, Benitez G, et al.. "Naked" gold nanoparticles supported on HOPG: Melanin functionalization and catalytic activity. Nanoscale. 2011;3(4):1708.

[136] Ju KY, Lee JW, Im GH, Lee S, Pyo J, Park SB, Lee JH, Lee JK. Bio-inspired, melanin-like nanoparticles as a highly efficient contrast agent for T1-weighted magnetic resonance imaging. Biomacromolecules. 2013;14(10):3491–3497.

[137] Seppä H. Superconducting transition-edge bolometer in a resistive and in an inductive mode. IEEE Trans Appl Supercond. 2001;11(1):759–761.

[138] Huang GS, Wang M Te, Su CW, Chen YS, Hong MY. Picogram detection of metal ions by melanin-sensitized piezoelectric sensor. Biosens Bioelectron. 2007;23(3):319–325.

[139] Gallas JM. Optical lens system incorporating melanin as an absorbing pigment for protection against electromagnetic radiation. Patent No. 4,698,374, 1987.

[140] Gallas JM. Medium incorporating melanin as an absorbing pigment for protection against electromagnetic radiation. Patent No. 5,047,447, 1991.

[141] Gallas J, Eisner M. Melanin polyvinyl alcohol plastic laminates for optical applications. Patent No. 7,029,758, 2006.

[142] Araújo M, Viveiros R, Correia TR, Correia IJ, Bonifácio VDB, Casimiro T, et al.. Natural melanin: A potential pH-responsive drug release device. Int J Pharm. 2014;469(1):140–145.

[143] Berliner DL, Erwin RL, McGee DR. Therapeutic uses of melanin. Patent No. 5,776,968, 1998.

[144] Berliner DL, Erwin RL, McGee DR. Methods of treating Parkinson's disease using melanin. Patent No. 5,210,076, 1993.

[145] Bettinger CJ, Bruggeman JP, Misra A, Borenstein JT, Langer R. Biocompatibility of biodegradable semiconducting melanin films for nerve tissue engineering. Biomaterials. 2009;30(17):3050–3057.

[146] Pezzella A, Barra M, Musto A, Navarra A, Alfè M, Manini P, et al.. Stem cell-compatible eumelanin biointerface fabricated by chemically controlled solid state polymerization. Mater Horiz. 2015;2(2):212–220.

[147] Rivnay J,Owens RM,Malliaras GG. The rise of organic bioelectronics. Chemistry of Materials. 2013;26(1):679-685.

[148] Solano F. Melanins: Skin pigments and much more—types, structural models, biological functions, and formation routes. New J Sci. 2014; 2014:1–28.

[149] Pawelek JM, Platt JT. Cosmetic melanins. Patent No. 5,744,125, 1998.

[150] D'Ischia M, Wakamatsu K, Napolitano A, Briganti S, Garcia-Borron JC, Kovacs D, et al.. Melanins and melanogenesis: Methods, standards, protocols. Pigment Cell Melanoma Res. 2013;26(5):616–633.

[151] Kunwar A, Adhikary B, Jayakumar S, Barik A, Chattopadhyay S,Raghukumar S, Priyadarsini KI. Melanin, a promising radioprotector: Mechanisms of actions in a mice model. Toxicol Appl Pharmacol. 2012;264(2):202–211.

[152] Zhang R, Fan Q, Yang M, Cheng K, Lu X, Zhang L, et al.. Engineering melanin nanoparticles as an efficient drug-delivery system for imaging-guided chemotherapy. Adv Mater. 2015;27(34):5063–5069.

[153] Riley PA. Melanin. Int J Biochem Cell Biol. 1997;29(Ii):1235–1239.

[154] Siegel SM. Fungal biosorption: A comparative study of metal uptake by penicillium and cladosporil'h. Met Speciat Sep Recover. 1987;1:339.

[155] Siegel SM, Galun M, Siegel BZ. Filamentous fungi as metal biosorbents: A review. Water Air Soil Pollut. 1990;53(3–4):335–344.

[156] Rizzo DM, Blanchette RA, Palmer MA. Biosorption of metal ions by *Armillaria rhizomorphs*. Can J Bot. 1992;70(8):1515–1520.

[157] Caporalin CB. Comparison of biosorption of rare earth metals by melanized biomass of *Aspergillus nidulans* fungus in free and immobilized forms [thesis]. Araraquara: Sao Paulo State University, 2011.

[158] Zhdanova NN, Vasilevskaya AI, Sadovnikov YS, Artyshkova LA. Dynamics of micromycete complexes from soils contaminated with radionuclides. Mikol i Fitopatol. 1990;24(6):504–512.

[159] Mahmoud YA. Uptake of radionuclides by some fungi. Mycobiology. 2004;32(3):110–114.

[160] Singleton I, Tobin JM. Fungal interactions with metals and radionuclides for environmental bioremediation. In: Frankland JC, Magan N, Gadd GM, editors. Fungi and Environmental Change. Cambridge: Cambridge University Press; 1996. p. 282–298.

[161] Steiner M, Linkov I, Yoshida S. The role of fungi in the transfer and cycling of radionuclides in forest ecosystems. J Environ Radioact. 2002;58(2):217–241.

[162] Zhdanova NN, Tugay T, Dighton J, Zheltonozhsky, V, Mcdermott P. Ionizing radiation attracts soil fungi. Mycol Res. 2004;108(9):1089–1096.

[163] Dixit R, Wasiullah, Malaviya D, Pandiyan K, Singh UB, Sahu A, et al.. Bioremediation of heavy metals from soil and aquatic environment: An overview of principles and criteria of fundamental processes. Sustain. 2015;7(2):2189–2212.

[164] Shakya M, Sharma P, Meryem SS, Mahmood Q, Kumar A. Heavy metal removal from industrial wastewater using fungi: Uptake mechanism and biochemical aspects. J Environ Eng. 2015; 142(9):C6015001.

Section 4

Melanin Analysis Methods

Chapter 5

Application of Electron Paramagnetic Resonance Spectroscopy to Examine Free Radicals in Melanin Polymers and the Human *Melanoma Malignum* Cells

Barbara Pilawa, Magdalena Zdybel and
Ewa Chodurek

Additional information is available at the end of the chapter

http://dx.doi.org/10.5772/67377

Abstract

The studies of free radicals in melanin and the human *melanoma malignum* cells by an X-band (9.3 GHz) electron paramagnetic resonance (EPR) spectroscopy were presented. The original results were compared with those published earlier. The aim of this work was application of the advanced spectral analysis to determine free radical properties in melanin biopolymers obtained from different melanotic tumor cells and free radicals existing in the human melanoma cells. Magnetic spin-lattice interactions in melanin samples were tested. The evolution of lineshape of tumor cells with increasing of microwave power was determined to confirm their complex free radical system. The useful shape parameters were proposed. The shape of melanotic tumor cells was analyzed. EPR spectra of free radicals in the melanin isolated from different tumor cells measured in the wide range of microwave power were analyzed. The melanins were obtained from the control tumor cells and the cells cultured with the several antitumor substances. The usefulness of the electron paramagnetic resonance spectroscopy was confirmed.

Keywords: melanin, human *melanoma malignum* cells, free radicals, paramagnetic centers, electron paramagnetic resonance spectroscopy

1. Introduction

Free radicals of natural melanin and the melanin in the human *melanoma malignum* cells were studied. Eumelanin and pheomelanin biopolymers are known [1–7]. Chemical structures of eumelanin and pheomelanin were shown in **Figure 1** [1]. Sulfur atoms exist in pheomelanin, but they were not found in eumelanin [1]. Eumelanin mainly exists in the human organism [1, 2]. Melanin was found in skin [3, 4], hair [5, 6], eye [7], and liver cells [8].

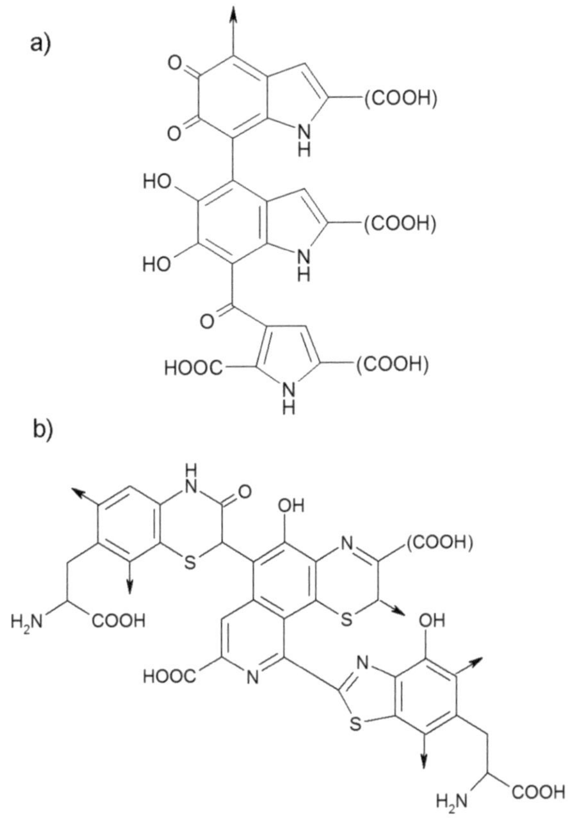

Figure 1. Chemical structure of eumelanin (a) and pheomelanin (b) [1].

Eumelanin was found in the melanotic tumor cells [9, 10]. Melanotic tumor cells are studied by NMR [11, 12], FTIR [13, 14], and HPLC [15, 16] methods. In this work we were interested in EPR studies of melanin from different tumor cells. Melanin polymers are known from paramagnetic character and o-semiquinone free radicals with spin S = 1/2 [17–40]. Unpaired electrons of free radicals obey the electron paramagnetic resonance (EPR) effect [17–40]. o-Semiquinone free radicals absorb microwaves in the magnetic field. This absorption is the base of electron paramagnetic resonance (EPR) spectroscopy [41–43]. Free radicals in eumelanin [17–19, 26, 40] and pheomelanin [22, 31, 38–40] are responsible for the EPR spectra, which differ in the shape. Typical EPR spectra of the model eumelanin DOPA-melanin and pheomelanin are shown in **Figure 2**. Comparison of the lineshape of these EPR spectra indicates that eumelanin reveals the simple single line (**Figure 2a**) and EPR line of pheomelanin reveals the complex shape with the unresolved hyperfine structure (**Figure 2b**). The lineshapes of the EPR spectra of DOPA-melanin [17–19, 26, 34–37] and pheomelanin [22, 31, 38–40] were presented in a lot of papers. EPR spectra were measured for free radicals in melanotic tumor cells [44–52].

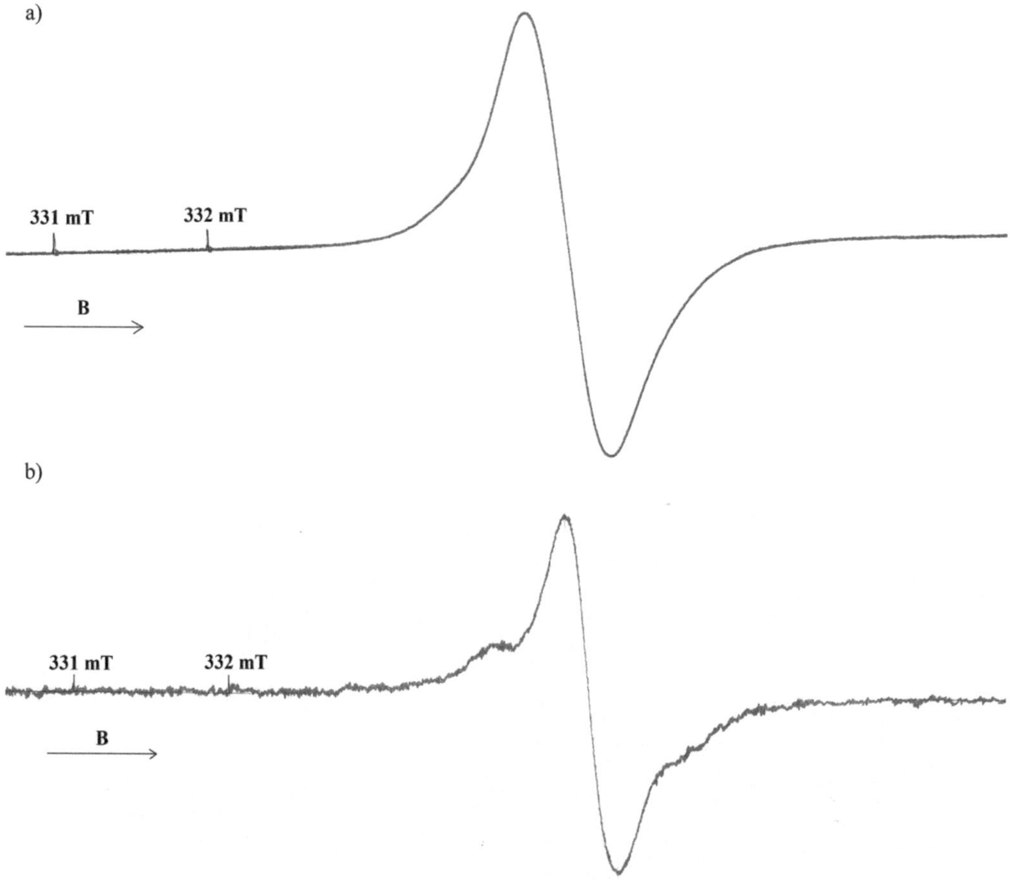

Figure 2. The EPR spectra of eumelanin (a) and pheomelanin (b). The measurements were done with the low microwave power with attenuation of 7 dB (microwave power of 14 mW). The melanin samples were studied in paper [40].

The drugs and substances with the antitumor interactions are still developed and searched [53, 54]. Melanin by free radicals interacts with drugs [55–57]. The effect of the antitumor substances on free radicals in the human tumor melanotic cells was shown by the use of the EPR spectroscopy [58–60].

The aim of this work was application of the advanced spectral analysis to determine free radical properties in melanin biopolymers obtained from the melanotic tumor cells and free radicals exiting in human melanoma cells. Free radicals in the original melanin samples and samples treated by the several antitumor substances were studied. The physical method of free radical detection based on paramagnetic character of melanins was used. EPR spectra of the tested natural melanins were compared with those of the model synthetic melanin polymers.

The innovatory lineshape analysis and the influence of microwave power on the complex EPR spectra were performed. The results were useful in medicinal therapy of the melanotic tumor cells. Both our published quantitative results [58–60] were cited, and the original spectral unpublished results were presented. The novelty in the present work, relative to our earlier papers [58–60], was the proposition of the spectral parameters to examine of the multicomponent EPR spectra as the sum of lines resulted from different types of free radicals existing in the melanotic A-2058 cells. The changes of these parameters with increasing of microwave power for the EPR spectra of the control cells and the cells cultured with valproic acid (VPA), 5,7-dimethoxycoumarin (DMC), and both valproic acid and 5,7-dimethoxycoumarin were presented.

2. Experimental

2.1. The tested antitumor substances

The influence of the following substances on human *melanoma malignum* cells, valproic acid (VPA) ($C_8H_{16}O_2$), 5,7-dimethoxycoumarin (DMC), and both VPA and DMC, was examined. Chemical structures of the tested substances are shown in **Figure 3** [61]. VPA and DMC were used as the potential antitumor substances [61].

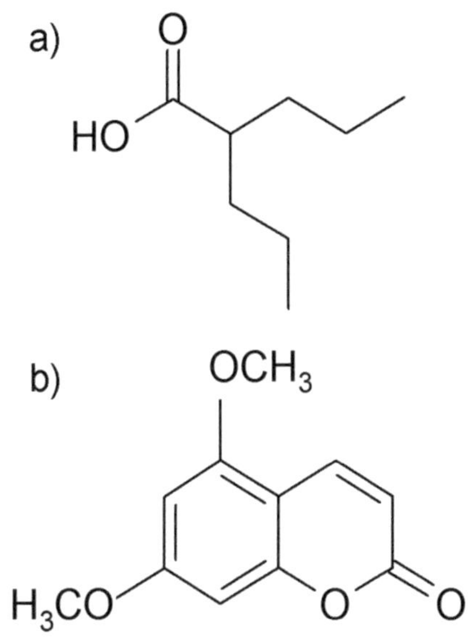

Figure 3. Chemical structures of valproic acid (VPA) (a) and 5,7-dimethoxycoumarin (DMC) (b) [61].

2.2. The tested human *melanoma malignum* cells

The three types of the human malignant melanoma cell lines, A-2058, A-375, and G-361, were used in this study. The cells were also cultured with the antitumor substances: valproic acid (VPA), 5,7-dimethoxycoumarin (DMC), and both VPA and DMC. In our EPR studies, the measurements were performed for the same number of cells.

The A-2058, A-375, and G-361 cells were obtained from LGC Promochem (Łomianki, Poland). A-2058 cells and A-375 cells were grown in the Minimum Essential Medium Eagle (Sigma-Aldrich). G-361 cells were grown in McCoy's medium (Sigma-Aldrich). These media were supplemented by the following components: 10% fetal bovine serum (FBS, PAA), 100 U/ml penicillin (Sigma-Aldrich), 100 μg/mL streptomycin (Sigma-Aldrich), and 10 mM HEPES (Sigma-Aldrich). The cells were incubated at temperature 37°C with the use of 5% CO_2. The incubation details were described in [58, 60].

The human malignant melanoma cell lines were incubated with 1 mM VPA, 10 μM DMC, and their combination for 4 days (A-2058) or 7 days (A-375 and G-361). EPR spectra of free radicals in the A-2058 cells and in melanin isolated from A-375, and G-361 cells were analyzed.

2.3. Isolation of melanin biopolymers from the melanotic cells

Melanin was isolated from the human *melanoma malignum* cells: A-375 and G-361. The enzymatic isolation procedure was described in detail in papers [62, 63]. The cells were lysed by incubation with 1% Triton X-100 (Sigma-Aldrich) for 1 hour at room temperature. The melanin was obtained by centrifugation of the lysates of the control cells, and the cells were cultured with VPA, DMC, and both VPA and DMC. The concentrations of VPA and DMC were 1 mM and 10 μM, respectively. The remaining pellets were washed with phosphate buffer, resuspended in Tris-HCl buffer (50 mM, pH 7.4), and incubated for 3 h at temperature 37°C. This Tris-HCl buffer contained sodium dodecyl sulfate (5 mg/ml) and proteinase K (0.33 mg/ml, Sigma-Aldrich). Melanin as the insoluble pigments was successively washed with 0.9% NaCl, methanol, and hexane, dried to a constant weight at temperature 37°C, and stored in a glass desiccator over P_2O_5.

2.4. The model eumelanin

The model eumelanin as DOPA-melanin was obtained by tyrosinase-catalyzed oxidation of 3,4-dihydroxyphenylalanine. The precursor (3,4-dihydroxyphenylalanine) was obtained from Sigma-Aldrich firm. The precursor was dissolved in 50 mM sodium phosphate buffer (pH 6.8). The final concentration was 2 mM. The reaction mixture after addition of tyrosinase (100 U/ml) was incubated for 48°C at temperature of 37°C. DOPA-melanin was obtained from the mixture by centrifugation (5000 × g, 15 min). The samples were washed by deionized water. Tyrosine was removed from melanin sample by treatment with SDS and methanol and NaCl. Finally, the sample was rewashed with deionized water and dried to a constant weight at temperature 37°C. This procedure was described in detail in [59, 60].

2.5. EPR measurements

2.5.1. EPR detection system

Free radicals in melanin biopolymers existing in different types of tumor cells and model synthetic melanin were examined by the use of electron paramagnetic resonance (EPR) spectroscopy. EPR spectra of melanin isolated from the cells and EPR spectra of the whole melanotic cells were tested. The first-derivative spectra were measured by an X-band (9.3 GHz) EPR spectrometer produced by Radiopan (Poznań, Poland) and the numerical data acquisition system—the Rapid Scan Unit of Jagmar (Kraków, Poland) (**Figure 4**).

Figure 4. The X-band (9.3 GHz) electron paramagnetic resonance (EPR) spectrometer of Radiopan (Poznań, Poland) and the numerical data acquisition system—the Rapid Scan Unit of Jagmar (Kraków, Poland).

The cells or melanin samples in thin-walled glass tubes were located in the resonance cavity in magnetic field produced by electromagnet of the EPR spectrometer (**Figure 5**). In the magnetic field, the Zeeman splitting appeared [41, 42]. Free radicals absorb microwaves according to the electron paramagnetic resonance condition [41, 42]:

$$h\nu = \mu_B g B_r \tag{1}$$

where h, Planck constant; ν, microwave frequency; μ_B, Bohr magneton; g-factor; and B_r, resonance magnetic induction.

The absorption is proportional to the free radical concentrations in the samples. The detailed determination of the free radical concentrations in cells and melanin samples was described in [58–60].

For the measurements and spectral analysis, the professional spectroscopic programs of Jagmar (Kraków, Poland), LabVIEW 8.5 of National Instruments (USA) and Origin (USA) were used. The Silesian Medical University has the right to use these programs. The program to spectroscopic analysis was prepared by Jagmar firm specially to our EPR spectrometer. The other programs are widely available.

Figure 5. The resonance cavity of the X-band (9.3 GHz) EPR spectrometer of Radiopan (Poznań, Poland).

2.5.2. The parameters of the EPR measurements

The EPR spectra were measured with the magnetic modulation of 100 kHz. Microwave frequency (ν) from the X-band (9.3 GHz) was obtained by MCM 101 detector of EPRAD (Poznań, Poland). The magnetic induction (B) in the range 332–338 mT was measured by NMR magnetometer of EPRAD (Poznań, Poland).

The maximal microwave power produced by klystron in microwave bridge of the EPR spectrometer was 70 mW. The measurements of the EPR spectra were done in the range of microwave power from 2.2 mW (attenuation of 15 dB) to 70 mW (attenuation of 0 dB). The microwave power was regulated by attenuation according to the formula [41, 42]:

$$\text{attenuation (dB)} = 10 \log M_o/M \tag{2}$$

where M is the microwave power used for detection of the EPR spectrum and M_o is the maximal microwave power (70 mW).

2.5.3. Analysis of the EPR spectra

The influence of microwave power in the range of 2.2–70 mW on the lineshape parameters of the EPR spectra of the tested samples was determined. The model first-derivative EPR spectrum with the values, A_1, A_2, B_1, and B_2, was shown in **Figure 6**. The lineshape parameters were obtained as A_1/A_2, A_1–A_2, B_1/B_2, and B_1–B_2.

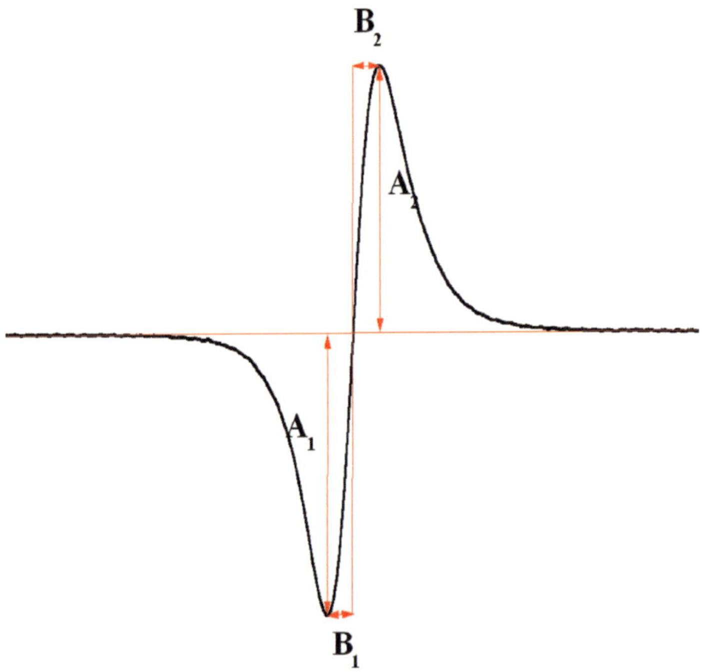

Figure 6. The exemplary model first-derivative EPR spectrum with the shape parameters: A_1, A_2, B_1, and B_2.

The evolution of the proposed lineshape parameter with increasing of microwave power gives information about complex free radical system in the biological samples.

The influence of microwave power on the integral intensities (*I*) of the EPR spectra was determined. Integral intensity (*I*) is proportional to the concentration of free radicals in the sample [41–43]. Integral intensity (*I*) of the EPR spectrum is the area under the absorption line [41–43]. Because the EPR spectra were measured as the first derivative of absorption, the spectral lines were double integrated to calculate the integral intensity. The first integration gives the absorption spectra. The second absorption gives the area under the absorption line.

The changes of integral intensity (*I*) of the EPR line with increasing of microwave power bring to light the spin-lattice interactions in the samples [41, 42]. Integral intensity (*I*) of the homogeneous broadened lines increased with increasing of microwave power, and after the reaching the maximal values, it decreased with the continuing increase of microwave power of the measurement [41, 42]. The faster spin-lattice relaxation caused microwave saturation of the EPR line at the higher microwave powers [41, 42].

3. Results and discussion

3.1. EPR spectra of free radicals in the human *melanoma malignum* A-2058 cells

Free radicals with the strong EPR lines of g-factor near 2 were found in A-2058 human melanoma cells [58]. The EPR spectra of the A-2058 cells recorded with the attenuation of microwave power of 7 dB were presented in **Figure 7**. The other spectra of these samples were presented in paper [58]. The EPR spectra are the broad nonsymmetrical lines (**Figure 7**). The broadening of the EPR lines of A-2058 cells is caused by dipolar interactions between free radicals. In this study we concentrated on the spin-lattice interactions in A-20058 cells and on their complex system of free radicals.

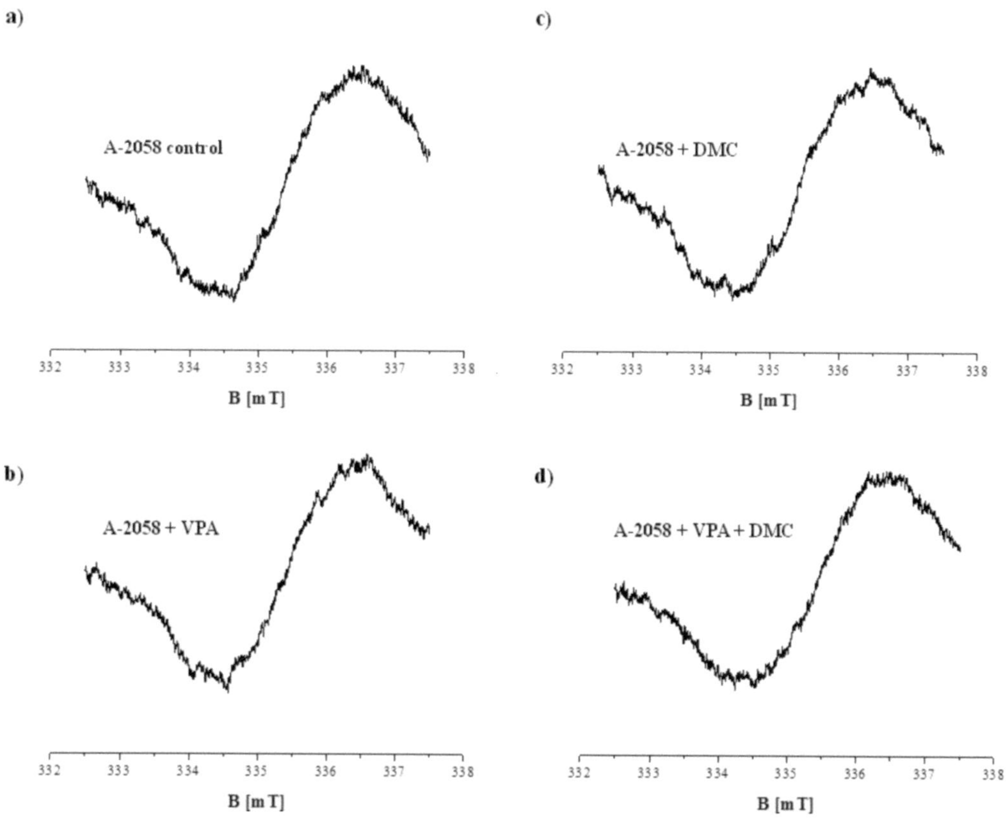

Figure 7. The EPR spectra of the human *melanoma malignum* control A-2058 cells (a) and the A-2058 cells cultured with VPA (b), DMC (c), and both VPA and DMC (d). The EPR spectra were measured with the attenuation of microwave power of 7 dB. *B*, magnetic induction.

The influence of the antitumor substances, VPA, DMC, and both VPA and DMC, on spin-lattice interactions in A-2058 human melanoma cells was determined. The influence of microwave power (M/M_o) on integral intensities (I) of the A-2058 cells for the control cells, and cells cultured with VPA, DMC, and both VPA and DMC, is compared in **Figure 8**.

Integral intensities (I) of the control A-2058 cells, the A-2058 cells cultured with VPA, and the A-2058 cells cultured with DMC increased with increasing microwave power (M/M_o). VPA and DMC did not change the character of changes of integral intensities (I) of A-2058 cells with microwave power (**Figure 8**). The absence of microwave saturation of the EPR lines of the control A-2058 cells, the A-2058 cells treated by VPA or treated by DMC in the microwave power up to 70 mW, indicated the fast spin-lattice relaxation processes existed in these cells.

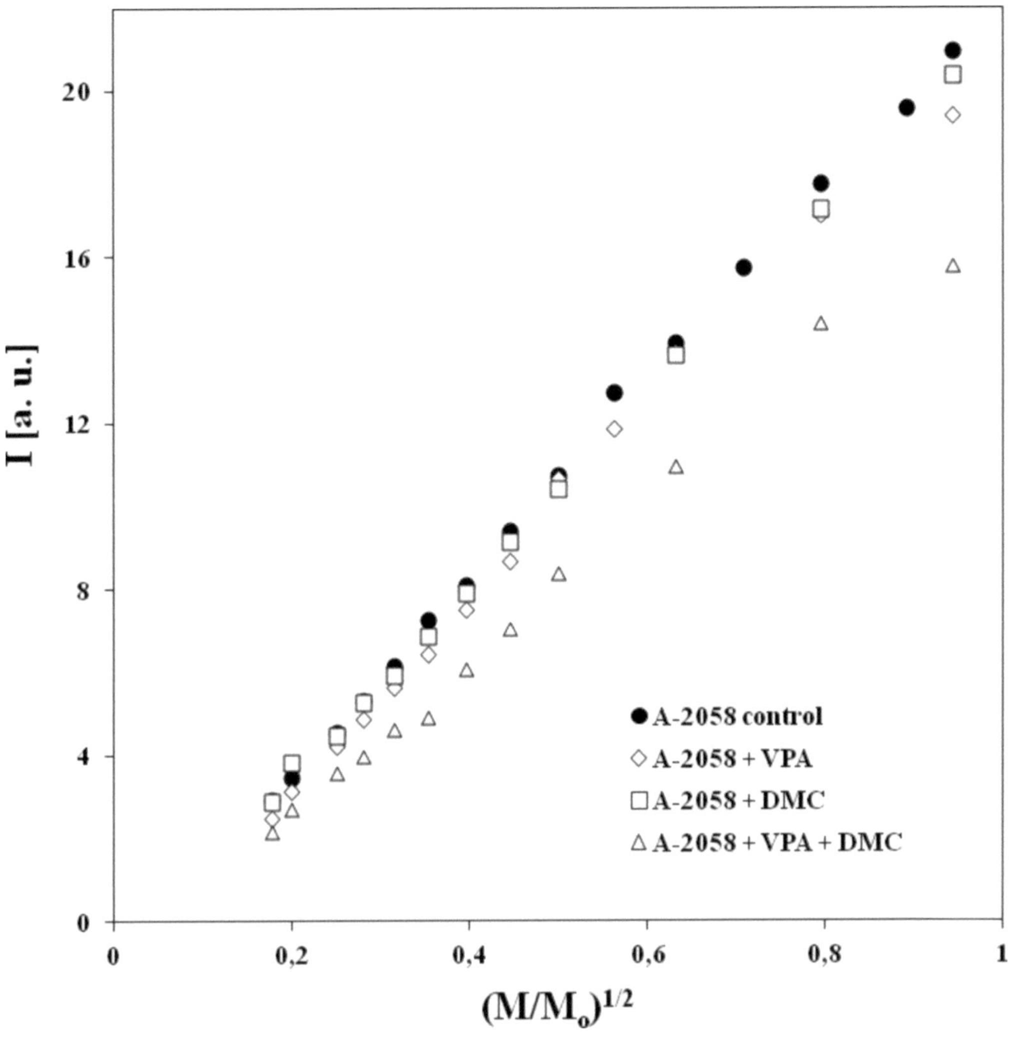

Figure 8. The influence of microwave power (M/M_o) on integral intensities (I) of the EPR spectra of the human *melanoma malignum* control A-2058 cells (•) and the A-2058 cells cultured with VPA (◊), DMC (□), and both VPA and DMC (Δ). M is the microwave power used during the measurement of the EPR spectrum, and M_o is the maximal microwave power produced by the klystron (70 mW).

The other situation was observed for the human malignant melanoma cell line A-2058 cultured with both VPA and DMC. The integral intensity (I) of the EPR lines of A-2058 cells treated with VPA and DMC together increased with increasing of microwave power (M/M_o) and it started saturating (**Figure 8**). The decrease of the integral intensity (I) was not observed, but the approaching to the maximum was visible (**Figure 8**). It means that the relatively slower spin-lattice relaxation processes existed in A-2058 cells cultured with both VPA and DMC, compared to the control cells, and the cells treated only with VPA or only with DMC. As one can see, the strongest effect on magnetic interactions in A-2058 cells was caused by the VPA and DMC used together in the cell culture.

The lineshape of the EPR spectra of the control A-2058 cells, and the A-2058 cells cultured with VPA, DMC, and both VPA and DMC, changed with increasing of microwave power (M/M_o). The changes of the analyzed lineshape parameters, A_1/A_2, A_1–A_2, B_1/B_2, and B_1–B_2, for the control A-2058 cells, and A-2058 cells cultured with VPA, DMC, and both VPA and DMC, with the increasing microwave power (M/M_o), were presented in **Figures 9–12**, respectively.

All the tested lineshape parameters (A_1/A_2, A_1–A_2, B_1/B_2, and B_1–B_2) for the control A-2058 cells and for the A-2058 cells cultured with the antitumor substances (VPA, DMC, and both VPA and DMC) were not constant, and their changes with microwave power were observed (**Figures 9–12**). The strongest changes of the parameters A_1–A_2 (**Figure 10**) and B_1–B_2 (**Figure 12**) were obtained. The changes of the spectral shape parameters with microwave power were not regular (**Figures 9–12**). These nonregular changes of the spectral shape parameters with microwave power confirmed the existence of several types of free radical in the tested A-2058 cells, both in the control cells and in the cells treated with the used antitumor substances.

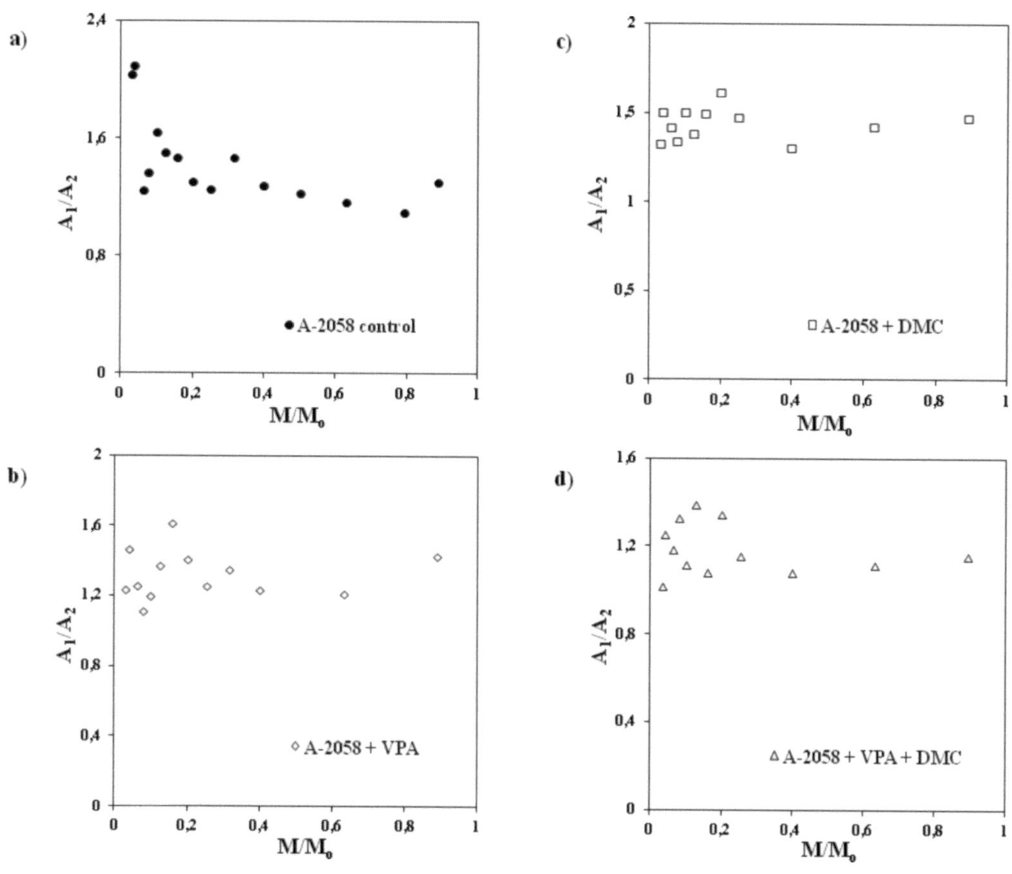

Figure 9. The influence of microwave power (M/M_o) on the lineshape parameter A_1/A_2, for the control A-2058 cells (•) (a) and A-2058 cells cultured with VPA (◊) (b), DMC (□) (c), and both VPA and DMC (Δ) (d). M is the microwave power used during the measurement of the EPR spectrum, and M_o is the maximal microwave power produced by the klystron (70 mW).

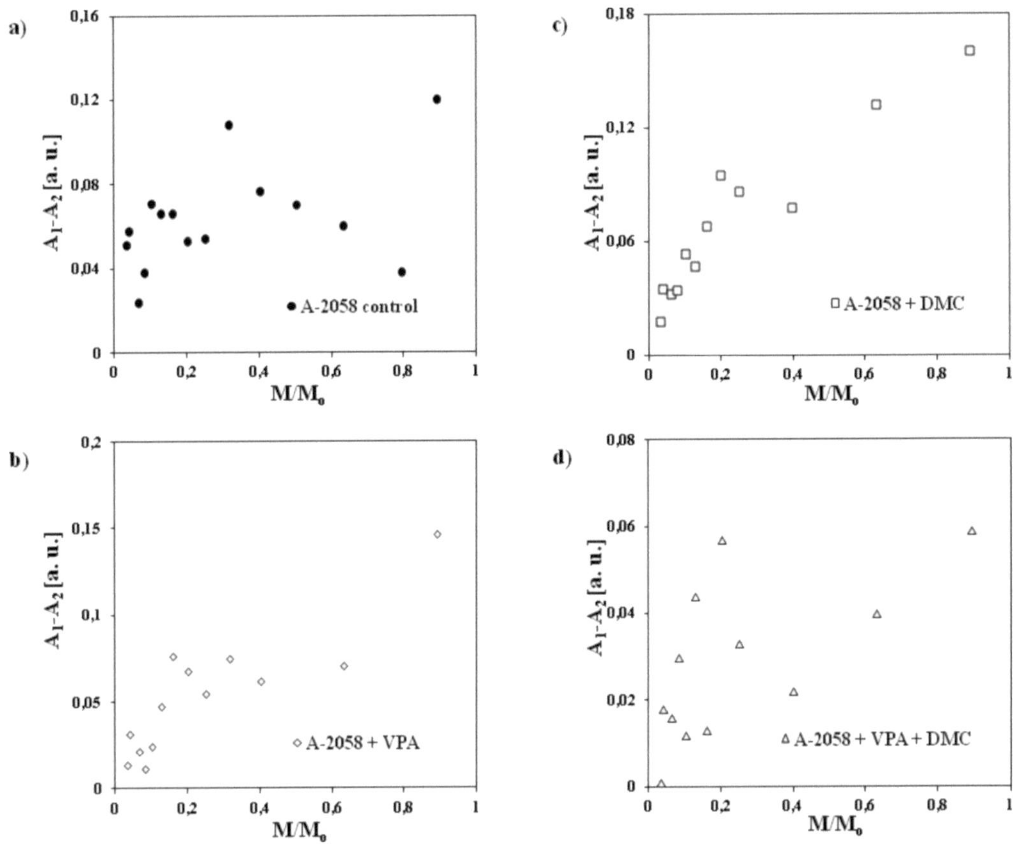

Figure 10. The influence of microwave power (M/M_0) on the lineshape parameter A_1–A_2, for the control A-2058 cells (•) (a) and A-2058 cells cultured with VPA (◊) (b), DMC (□) (c), and both VPA and DMC (Δ) (d). *M* is the microwave power used during the measurement of the EPR spectrum, and M_0 is the maximal microwave power produced by the klystron (70 mW).

We proposed these shape parameters, A_1/A_2, A_1–A_2, B_1/B_2, and B_1–B_2, for checking the multi-component type of free radical in cells. They supported in the analysis of complex free radicals in the other paramagnetic samples, for example, for drugs [64, 65]. The EPR spectra of the cells were superposition of several lines resulted from the individual groups of free radicals. The microwave power differently influenced these EPR components, dependent on the type of free radicals. Amplitudes (A), linewidths (ΔB_{pp}), and integral intensities (*I*) of each component lines changed differently with microwave power. The component EPR lines saturated at different microwave powers. All these facts resulted in the summary effects of nonregular changes of shape parameters with microwave power used during the measurements of the EPR spectra of A-2058 cells. The existence of several groups of free radicals in A-2058 cells was expected. The o-semiquinone free radicals, biradicals, and free radicals formed, for example, by UV irradiation of the cells, may exist in the A-2058 cells. The studies of the complex free radicals system in tumor cells with application of the spectral shape analysis in the broad range of microwave power will be continued. The numerical analysis of the components will be performed.

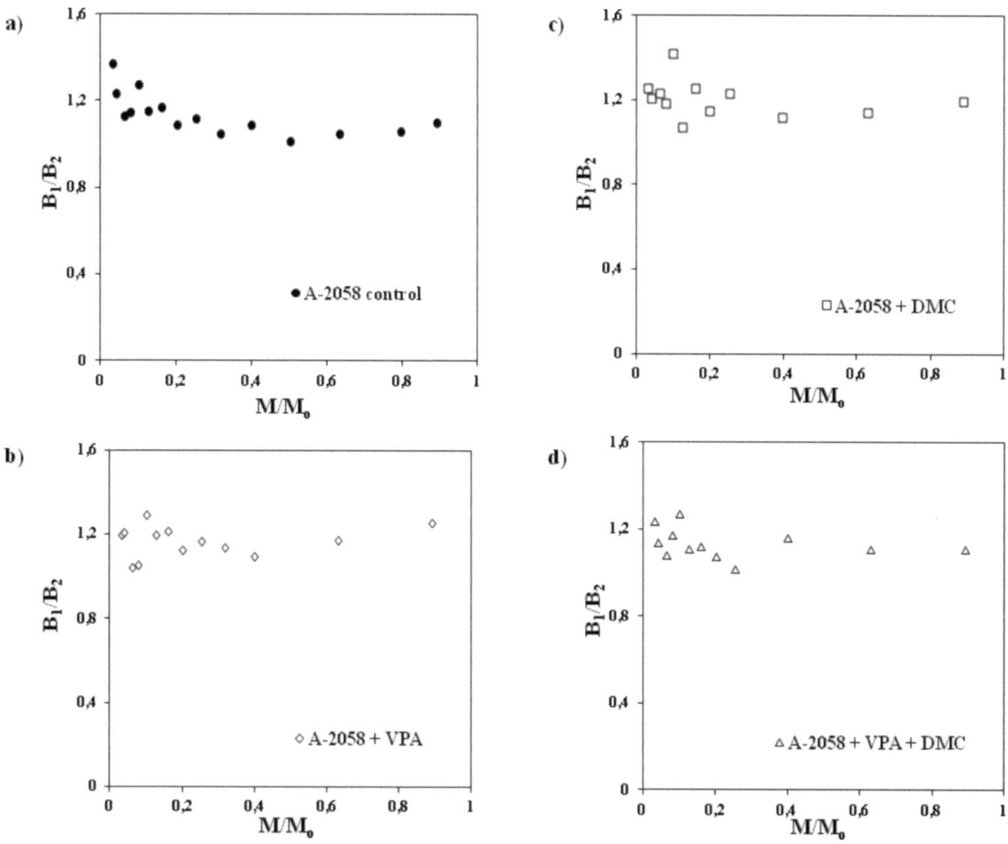

Figure 11. The influence of microwave power (M/M_0) on the lineshape parameter B_1/B_2, for the control A-2058 cells (•) (a) and A-2058 cells cultured with VPA (◊) (b), DMC (□) (c), and both VPA and DMC (Δ) (d). *M* is the microwave power used during the measurement of the EPR spectrum, and M_0 is the maximal microwave power produced by the klystron (70 mW).

Besides the shape analysis proposed in this work, the important qualitative results for free radicals in the human *melanoma malignum* A-2058 cells were obtained by us earlier [58]. It was pointed out that treatment by VPA, DMC, and both VPA and DMC decreased free radical concentration in A-2058 cells [58]. This effect was the strongest for VPA used together with DMC, so these substances were proposed as the antitumor drugs [58]. The used in the present work spectral parameter - integral intensity (*I*) - was more precise than the amplitude (*A*) [58] for examination of spin-lattice relaxation processes in A-2058 human melanoma cells.

3.2. EPR spectra of free radicals in melanin isolated from human *melanoma malignum* A-375 cells

Free radicals were also found in melanin biopolymer isolated from the control A-375 cells and the A-375 cells cultured with VPA, DMC, and both VPA and DMC. For all the melanin samples, EPR spectra were measured. The exemplary EPR spectra of melanin isolated from

A-375 cells cultured with VPA and DMC, recorded with microwave power attenuation of 7 dB, were shown in **Figure 13**. The other EPR spectra of melanin originated from A-375 cells were shown in [60].

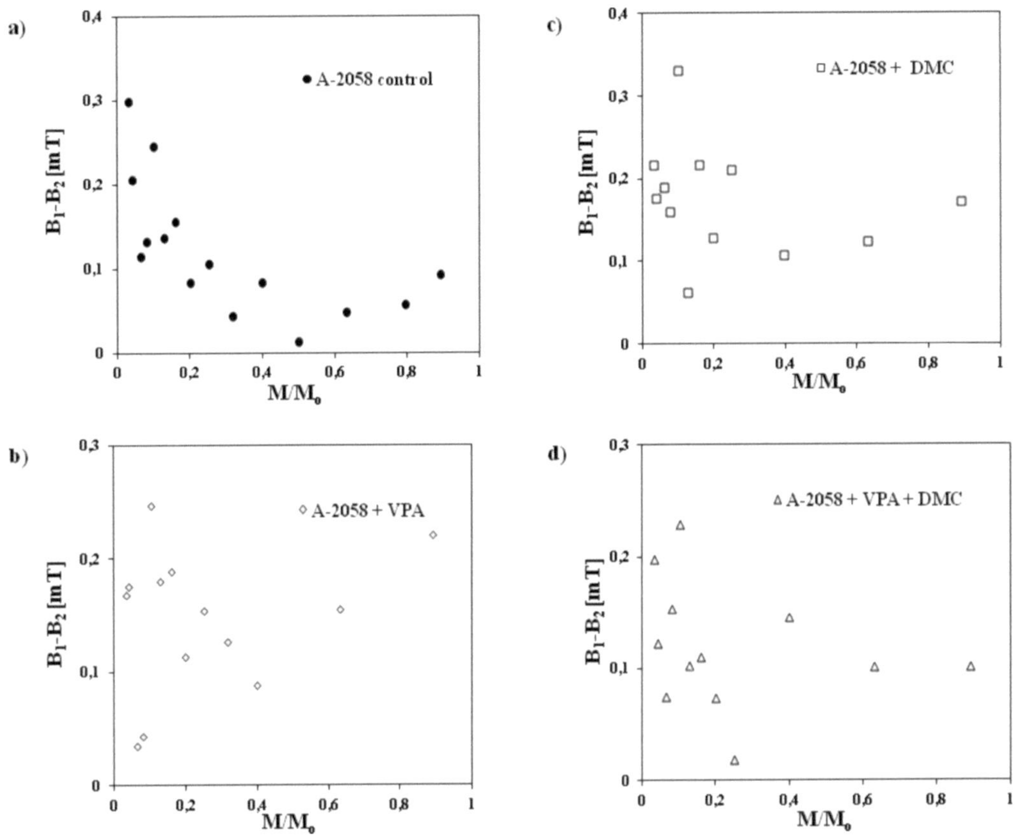

Figure 12. The influence of microwave power (M/M_0) on the lineshape parameter B_1–B_2, for the control A-2058 cells (•) (a) and A-2058 cells cultured with VPA (◊) (b), DMC (□) (c), and both VPA and DMC (Δ) (d). M is the microwave power used during the measurement of the EPR spectrum, and M_0 is the maximal microwave power produced by the klystron (70 mW).

The parameters of the EPR spectra of the melanin obtained from A-375 cells changed with microwave power. In **Figure 14**, the influence of microwave power on integral intensities (*I*) of the melanin obtained from A-375 cells cultured with VPA, DMC, and both VPA and DMC was compared. The changes of the integral intensities (*I*) of the melanin isolated from the control A-375 cells and the other A-375 cell culture with VPA, with increasing of microwave power, were published in our earlier paper [59].

The integral intensities (*I*) of the EPR lines of melanin isolated from A-375 cells treated with VPA increased with increasing of microwave power (M/M_0) reached the maximum and started to saturate (**Figure 14**). The EPR lines of melanin isolated from the control A-375 cells saturated at the low microwave power [59]. Comparing the results for EPR lines of melanin from the control A-375 cells [59] and from the A-375 cells cultured with VPA (**Figure 14**), it may be concluded that the faster spin-lattice relaxation processes existed in melanin from the A-375 cells treated by VPA. Such effect was not observed for the melanin isolated from A-375 cells cultured with DMC. EPR lines of melanin from A-375 cells treated with DMC (**Figure 14**) saturated at similar microwave power than the lines of melanin from the control A-375 cells [59]. The EPR lines of melanin obtained from A-375 cells treated by both VPA and DMC (**Figure 14**) saturated at the lower microwave power than the EPR lines of the melanin isolated from control cells [59]. The slower spin-lattice relaxation processes existed in melanin from A-375 cells cultured with both VPA and DMC than the EPR lines of the melanin from the control cells.

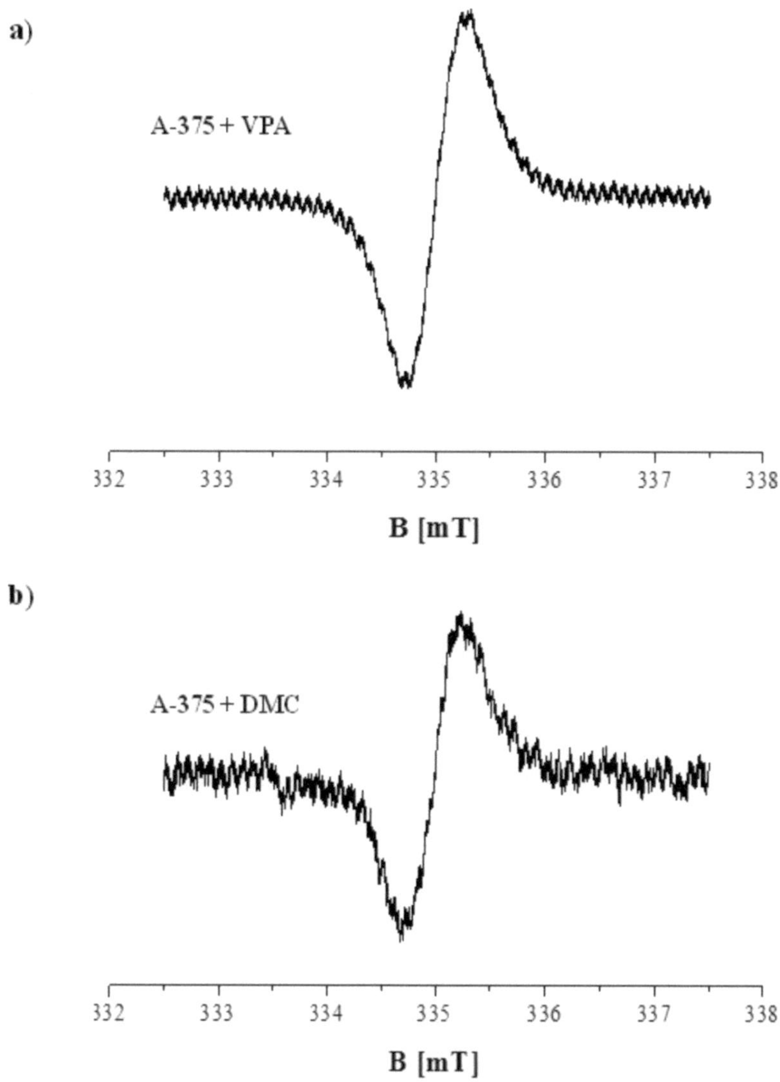

Figure 13. The EPR spectra of melanin isolated from the human *melanoma malignum* A-375 cells cultured with VPA (a) and DMC (b). The EPR spectra were measured with the attenuation of microwave power of 7 dB. *B*, Magnetic induction.

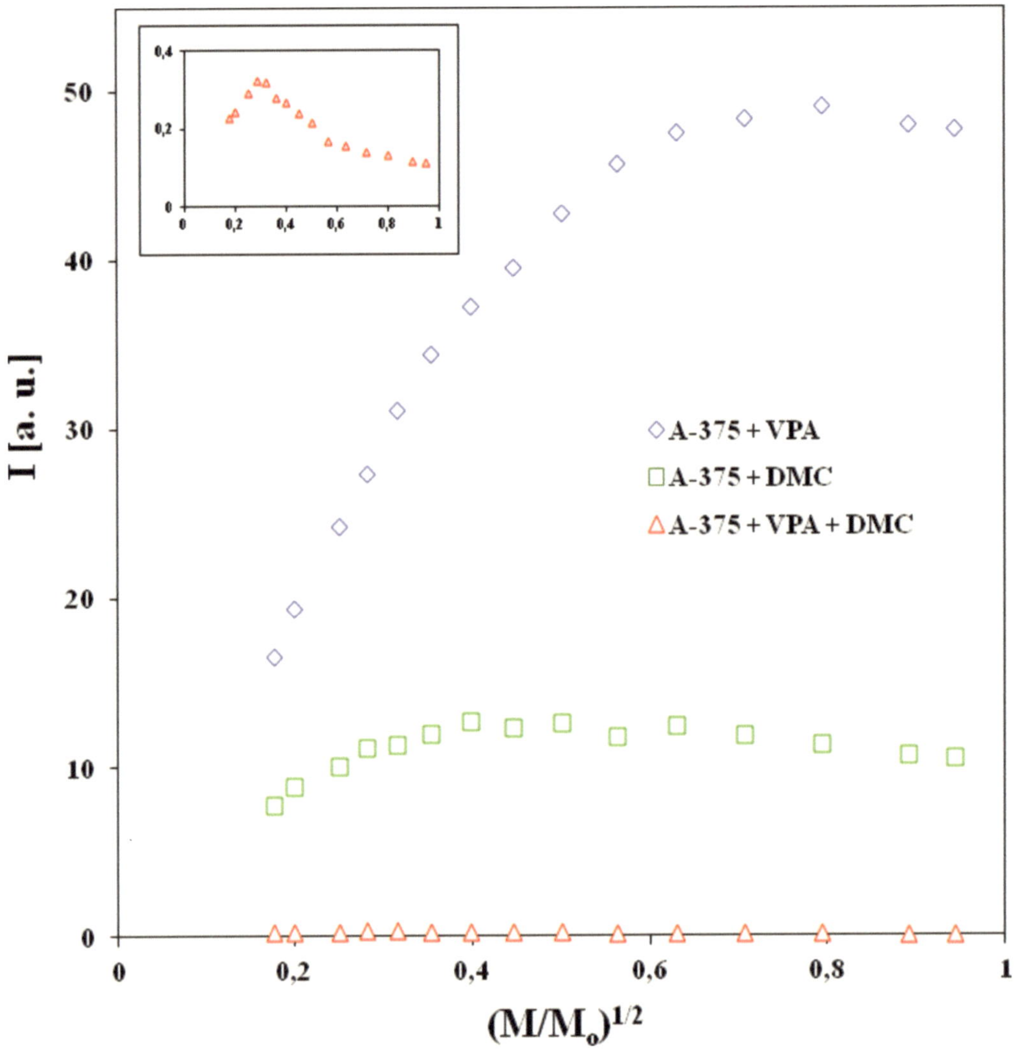

Figure 14. The influence of microwave power (M/M_o) on integral intensities (I) of the EPR spectra of melanin isolated from the human *melanoma malignum* A-375 cells cultured with VPA (◊), DMC (□), and both VPA and DMC (Δ). M is the microwave power used during the measurement of the EPR spectrum, and M_o is the maximal microwave power produced by the klystron (70 mW).

o-Semiquinone free radicals mainly existed in the melanin samples from A-375 cells. The quantitative results were published in the earlier paper [59, 60]. Considerable decrease of free radical concentration in melanin after treatment A-375 cells by both VPA and DMC was observed [60]. Free radical concentration in melanin isolated from A-375 cells cultured with DMC was lower than in melanin from the cells cultured with VPA [60]. The changes of amplitudes (A) and linewidths (ΔB_{pp}) with microwave power indicated homogeneous broadening of the EPR lines of melanin isolated from A-375 cells [60].

http://dx.doi.org/10.5772/67377

3.3. EPR spectra of free radicals in melanin isolated from human *melanoma malignum* G-361 cells

EPR lines of o-semiquinone free radicals were also measured for melanin isolated from G-361 human melanoma cells. The EPR spectra of melanin isolated from the control G-361 cells, and the G-361 cells treated with VPA, DMC, and both VPA and DMC, measured with microwave power attenuation of 7 dB, were shown in **Figure 15**. The other spectra of these melanin samples were presented in paper [60]. The high level of the noise was visible in these spectra (**Figure 15**), so the lower contents of free radicals were found in melanin from G-361 cells than from A-375 cells (**Figure 13**).

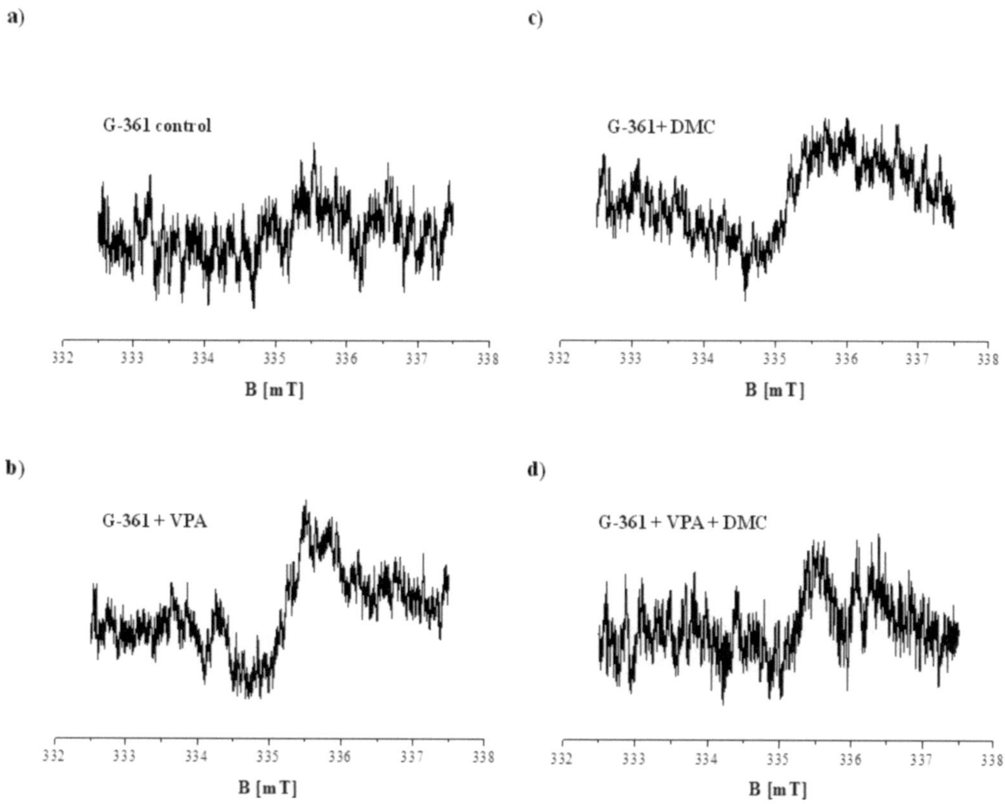

Figure 15. The EPR spectra of melanin isolated from the human *melanoma malignum* control G-361 cells (a), and the G-361 cells cultured with VPA (b), DMC (c), and both VPA and DMC (d). The EPR spectra were measured with the attenuation of microwave power of 7 dB. *B*, Magnetic induction.

The influence of the antitumor substances, VPA, DMC, and both VPA and DMC, on spin-lattice interactions in melanin obtained from G-361 human melanoma cells was not stated. The changes of integral intensities (*I*) of the melanin from G-361 cells for the control cells, and cells cultured with VPA, DMC, and both VPA and DMC, with increasing of microwave power

(M/M_0), were compared in **Figure 16**. The similar correlations between integral intensity (I) and microwave power for all the melanin samples were visible (**Figure 16**). The antitumor drugs did not change magnetic interactions in melanin structures of G-361 cells.

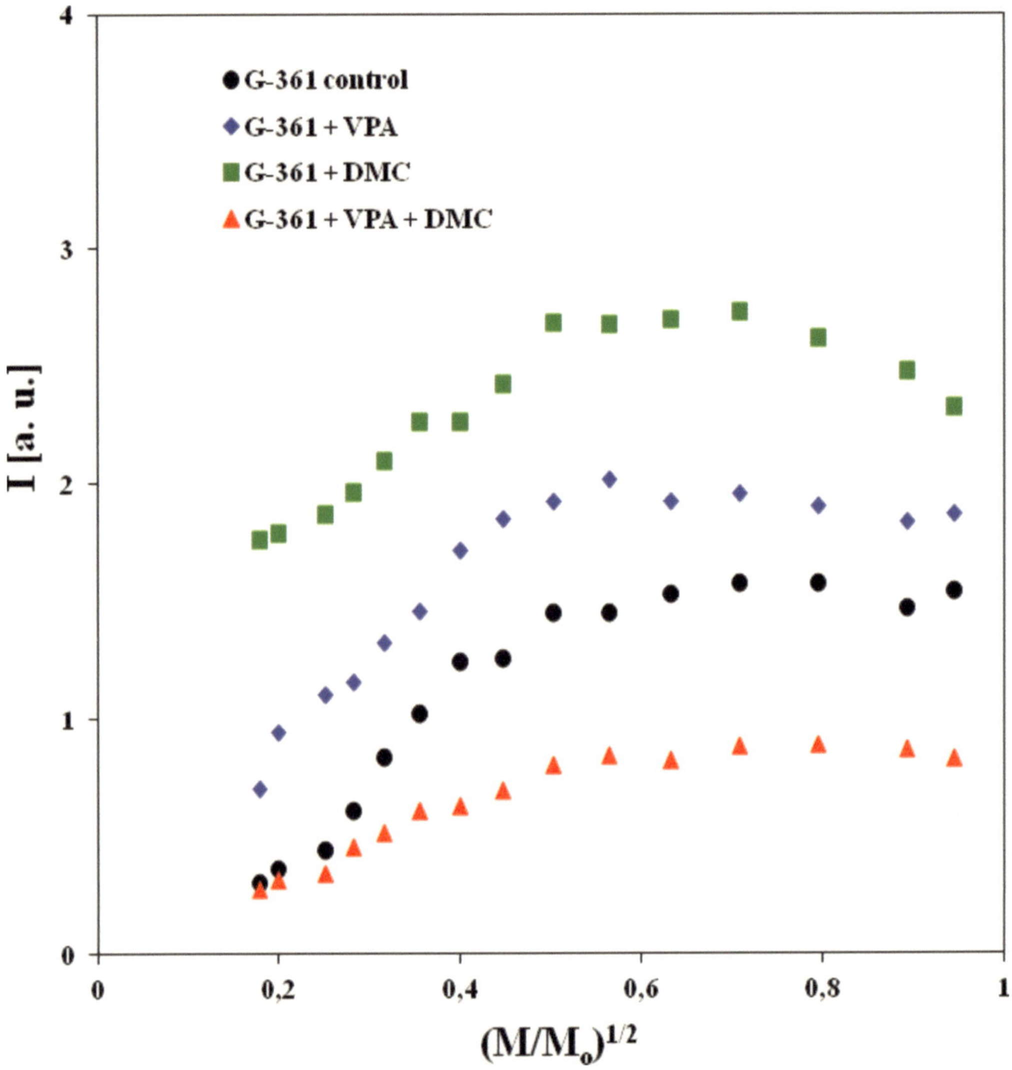

Figure 16. The influence of microwave power (M/M_0) on integral intensities (I) of the EPR spectra of melanin isolated from the human *melanoma malignum* control G-361 cells (•) and the G-361 cells cultured with VPA (♦), DMC (■), and both VPA and DMC (▲). M is the microwave power used during the measurement of the EPR spectrum, and M_0 is the maximal microwave power produced by the klystron (70 mW).

The quantitative results of EPR examination of melanin originated from G-361 cells were described in paper [60]. It was obtained that after treating of G-361 cells with both VPA and DMC free radical concentration in melanin strongly decreased [60]. Free radical concentration in melanin isolated from G-361 cells cultured with DMC was higher than in melanin from the cells cultured with VPA [60]. The changes of amplitudes (A) and linewidths (ΔB_{pp}) with microwave power indicated homogeneous broadening of the EPR lines of melanin isolated from G-361 cells [60]. Our present spin-lattice relaxation studies by the use of integral intensities (I) dependence on microwave power confirmed the results obtained for melanin from G-361 cells from the amplitude (A) changes with microwave power [60].

4. Conclusions

The existence of o-semiquinone free radicals in melanin from the human *melanoma malignum* cells was confirmed. Free radicals of melanin were mainly responsible for the EPR lines of the tested tumor cells. The free radical concentrations depended on the type of tumor cells. The antitumor drugs changed the free radical concentrations. The changes depended on the drug amounts. The parameters and lineshape of the EPR spectra of melanin changed with increasing of the measuring microwave power. All the EPR lines of the tested melanins were very broad. The most of the spin-lattice relaxation processes in melanin samples characterized the long relaxation times, and their EPR lines saturated at the low microwave powers. The analysis of the lineshape of the EPR spectra measured in the wide range of microwave power was useful to obtain information about complex free radical system in the melanin biopolymers. The spectral EPR results may be applied in therapy of tumors contained melanin. The free radical concentrations in the tumors and the effect of the antitumor substances on their values may be obtained. The effective antitumor drugs as those which cause the decrease of free radical concentrations in the melanotic tumor cells may be spectroscopically found.

Acknowledgments

The presented electron paramagnetic resonance examinations of free radicals in melanin and tumor cells were supported by Medical University of Silesia, Katowice, Poland (grant no. KNW-1-005/K/6/0).

Author details

Barbara Pilawa[1], Magdalena Zdybel[1]* and Ewa Chodurek[2]

*Address all correspondence to: mzdybel@sum.edu.pl

1 Department of Biophysics, School of Pharmacy with the Division of Laboratory Medicine in Sosnowiec, Medical University of Silesia, Katowice, Poland

2 Department of Biopharmacy, School of Pharmacy with the Division of Laboratory Medicine in Sosnowiec, Medical University of Silesia, Katowice, Poland

References

[1] Wakamatsu K, Ito S. Advanced chemical methods in melanin determination. Pigment Cell & Melanoma Research. 2002;**15(3)**:174-183. DOI: 10.1034/j.1600-0749.2002.02017.x

[2] Ito S, Wakamatsu K. Quantitative analysis of eumelanin and pheomelanin in humans, mice and other animals: a comparative review. Pigment Cell & Melanoma Research. 2003;**16(5)**:523-531. DOI: 10.1034/j.1600-0749.2003.00072.x

[3] Thody AJ, Higgins EM, Wakamatsu K, Ito S, Burchill SA, Marks J. Pheomelanin as well as eumelanin is present in human epidermis. Journal of Investigative Dermatology. 1991;**97(2)**:340-344. DOI: 10.1111/1523-1747.ep12480680

[4] Hennessy A, Oh C, Diffey B, Wakamatsu K, Ito S, Rees J. Eumelanin and pheomelanin concentration in human epidermis before and after UVB irradiation. Pigment Cell & Melanoma Research. 2005;**18(3)**:220-223. DOI: 10.1111/j.1600-0749.2005.00233.x

[5] Ito S, Fujita K. Microanalysis of eumelanin and pheomelanin in hair and melanomas by chemical degradation and liquid chromatography. Analytical Biochemistry. 1985;**144(2)**:527-536. DOI: 10.1016/0003-2697(85)90150-2

[6] Ozeki H, Ito S, Wakamatsu K, Thody A. Spectrophotometric characterization of eumelanin and pheomelanin in hair. Pigment Cell & Melanoma Research. 1996;**9(5)**:265-270. DOI: 10.1111/j.1600-0749.1996.tb00116.x

[7] Hu DN, Simon JD, Sarna T. Role of ocular melanin in ophthalmic physiology and pathology. Photochemistry and Photobiology. 2008;**84(3)**:639-644. DOI: 10.1111/j.1751-1097.2008.00316.x

[8] Gallone A, Sagliano A, Guida G, Ito K, Wakamatsu K, Capozzi V, Perna G, Zanna P, Cicero R. The melanogenic system of the liver pigmented macrophages of *Rana esculenta* L. – Tyrosinase activity. Histology and Histopathology. 2007;**22**:1065-1075.

[9] Klein-Szanto A, Bradl M, Porter S, Mintz B. Melanosis and associated tumors in transgenic mice. Proceeding of the National Academy of Sciences of the USA. 1991;**88**:169-173.

[10] Gowda KK, Agrawal P, Vaiphei K. Clear cell renal cell carcinoma with melanocytic differentiation: a case report, with review of the literature. Austin Journal of Clinical Pathology. 2014;**1(3)**:1-3.

[11] Bines SD, Tomasovic SP, Frazer JW, Leveque J, Boddie AW, Dennis L. Proton NMR examination of tumor cells of high or low metastatic potential. Clinical & Experimental Matastasis. 1987;**5(3)**:259-274. DOI: 10.1007/BF00124307

[12] Griffin JL, Shockcor JP. Metabolic profiles of cancer cells. Nature Reviews (Cancer). 2004;**4**:551-561. DOI:10.1038/nrc1390

[13] Bellisola G, Sorio C. Infrared spectroscopy and microscopy in cancer research and diagnosis. American Journal of Cancer Research. 2012;**2(1)**:1-21.

[14] Derenne A, Van Hemelryck V, Lamoral-Thevs D, Kiss R, Goormaghtigh E. FTIR spectroscopy: a new valuable tool to classify the effects of polyphenolic compounds on cancer cells. Biochimica et Biophysica Acta (BBA) – Molecular Basis of Disease. 2013;**1832(1)**:46-56. DOI: 10.1016/j.bbadis.2012.10.010

[15] Torabi-Pour N, Nouri AM, Chinequwndo F, Olivier RT. Standardization and potential use of HPLC for detection of cellular placental alkaline phosphatase using established tumor cell lines and fresh tumor biopsies. Urologia Internationalis. 1999;**63(4)**:234-241. DOI:10.1159/000030457

[16] Barbero MC, Garcia M, Almena A, Alino SF. HPLC analysis of supernatant soluble factors from tumor cell variants with defined metastatic status. Cellular and Molecular Biology. 1989;**35(4)**:441-447.

[17] Sarna T. Examination of structure and properties of melanin active centers. Zagadnienia Biofizyki Współczesnej. 1981;**6**:201-219 (in Polish).

[18] Pasenkiewicz-Gierula M, Sealy RC. Analysis of the ESR spectrum of synthetic dopa melanin. Biochimica et Biophysica Acta (BBA) - General Subjects. 1986;**884**:510-516. DOI: 10.1016/0304-4165(86)90202-3

[19] Sarna T, Plonka PM. Biophysical studies of melanin: paramagnetic, ion-exchange and redox properties of melanin pigments and their photoreactivity. In: Eaton SR, Eaton GR, Berliner LJ, editors. Biological Magnetic Resonance. Vol. 23, Biomedical EPR, Part A: Free Radicals, Metals, Medicine and Physiology. New York, Boston, Dordrecht, London, Moscow: Kluwer Academic/Plenum Publishers; 2005. pp. 125-142.

[20] Strzelczyk G, Sterniczuk M, Sadło J, Kowalska M, Michalik M. EPR study of γ-irradiated feather keratin and human fingernails concerning retrospective dose assessment. Nuleonika. 2013;**58(4)**:505-509.

[21] Almeida-Paes R, Frases S, de Sousa Araujo G, de Oliveira MME, Gerfen GJ, Nosanchuk JD, Zancope-Oliveira RM. Biosynthesis and functions of a melanoid pigment produced by species of the *Sporothrix* complex in the presence of L-tyrosine. Applied and Environmental Microbiology. 2012;**78(24)**:8623-8630. DOI: 10.1128/AEM.02414-12

[22] Chikvaidze EN, Partskhaladze TM, Gogoladze TV. Electron spin resonance (ESR/EPR) of free radicals observed in human red hair: a new, simple empirical method of determination of pheomelanin/eumelanin ration in hair. Magnetic Resonance in Chemistry. 2014;**52(7)**,377-382. DOI:10.1002/mrc.4075

[23] Mostert AB, Hanson GR, Sarna T, Gentle JR, Powell BJ, Meredith P. Hydration-controlled X-band EPR spectroscopy: a tool for unravelling the complexities of the solid-state free radical in eumelanin. Journal of Physical Chemistry B. 2013;**117(17)**:4965-4972. DOI: 10.1021/jp401615e

[24] Pal AK, Gaijar DU, Vasavada AR. DOPA and DHN pathway orchestrate melanin synthesis in *Aspergillus* species. Medical Mycology. 2014;**52(1)**:10-18. DOI: 10.3109/13693786.2013.826879

[25] Arun G, Eyini M, Gunasekaran P. Characterization and biological activities of extracellular melanin produced by *Schizophyllum commune* (Fries). Indian Journal of Experimental Biology. 2015;**53**:380-387.

[26] Costa TG, Szpoganicz B, Caramori GF, de Almeida VR, Mangrich AS, Mangoni AP. Spectroscopy and theoretical studies of natural melanin (eumelanin) and its complexation by iron(III). Journal of Coordination Chemistry. 2014;**67(6)**:986-1001. DOI: 10.1080/00958972.2014.905686

[27] Sajjan S, Kulkarni G, Yaligara V, Kyoung L, Karegoudar TB. Purification and physiochemical characterization of melanin pigment from *Klebsiella* sp. GSK. Journal of Microbiology and Biotechnology. 2010;**20(11)**:1513-1520. DOI: 10.4014/jmb.1002.02006

[28] Prados-Rosales R, Toriola S, Nakouzi A, Chatterjee S, Stark R, Gerfen G, Tumpowsky P, Dadachova E, Casadevall A. Structural characterization of melanin pigments from commercial preparations of the edible mushroom *Auricularia auricular*. Journal of Agricultural and Food Chemistry. 2015;**63(33)**:7326-7332. DOI: 10.1021/acs.jafc.5b02713

[29] Plonka PM. Electron paramagnetic resonance as a unique tool for skin and hair research. Experimental Dermatology. 2009;**18(5)**:472-484. DOI: 10.1111/j.1600-0625.2009.00883.x

[30] Plonka PM, Michalczyk D, Popik M, Handjiski B, Paus R. Electron paramagnetic resonance (EPR) spectroscopy for investigating murine telogen skin after spontaneous or depilation-induced hair growth. Journal of Dermatological Science. 2008;**49(3)**:227-240. DOI: 10.1016/j.jdermsci.2007.09.010

[31] Wolnicka-Glubisz A, Pecio A, Podkowa D, Kolodziejczyk LM, Plonka PM. Pheomelanin in the skin of *Hymenochirus boettgeri* (Amphibia: Anura: Pipidae). Experimental Dermatology. 2012;**21(7)**:537-540. DOI: 10.1111/j.1600-0625.2012.01511.x

[32] Michalczyk-Wetula D, Wieczorek D, Płonka PM. Splenic melanosis in agouti and black mice. Acta Biochimica Polonica. 2015;**62(3)**:457-463.

[33] Surwase SN, Jadhav SB, Phugare SS, Jadhav JP. Optimization of melanin production by *Brevundimonas* sp. SGJ using response surface methodology. 3 Biotech. 2013;**3(3)**:187-194. DOI: 10.1007/s13205-012-0082-4

[34] Bilińska B, Pilawa B, Zawada Z, Wylegala E, Wilczok T, Dontsov AE, Sakina NL, Ostrovsky MA, Ilyasova VB. Electron spin resonance investigations of human retinal pigment epithelium melanosomes from young and old donors. Spectrochimica Acta Part A. 2002;**58**:2257-2264.

[35] Otręba M, Zdybel M, Pilawa B, Beberok A, Wrześniok D, Rok J, Buszman E. EPR spectroscopy of chlorpromazine-induced free radical formation in normal human melanocytes. European Biophysical Journal. 2015;**44(5)**:359-365. DOI: 10.1007/s00249-015-1029-6

[36] Najder-Kozdrowska L, Pilawa B, Buszman E, Wieckowski AB, Świątkowska L, Wrześniok D, Wojtowicz W. Triplet states in DOPA-melanin and its complexes with kanamycin and copper Cu (II) ions. Acta Physica Polonica A. 2010;**118(4)**:613-618.

[37] Zdybel M, Chodurek E, Pilawa B. Free radicals in ultraviolet irradiated melanins and melanin complexes with Cd(II) and Cu(II) – EPR examination. Journal of Applied Biomedicine. 2015;**13(2)**:131-141. DOI: 10.1016/j.jab.2015.01.002

[38] Zdybel M, Pilawa B. Effect of UV irradiation on free radicals in synthetic melanin and melanin biopolymer from *Sepia officinalis* – EPR examination. Nukleonika. 2015;**60(3)**:483-488. DOI: 10.1515/nuka-2015-0085

[39] Pilawa B, Zdybel M, Buszman E, Witoszyńska T, Brotoń B. Paramagnetism of pigmented soil fungi of *Cladosporium herbarum*. Engineering of Biomaterials. 2009;**12(89–91)**:172-174.

[40] Matuszczyk M, Buszman E, Pilawa B, Witoszyńska T, Wilczok T. Cd^{2+} effect on free radicals in *Cladosporium cladosporioides* – melanin tested by EPR spectroscopy. Chemical Physics Letters. 2004;**394**:366-371.

[41] Wertz JE, Bolton JR. Electron spin resonance: elementary theory and practical applications. New York: Chapman and Hall; 1986.

[42] Stankowski J, Hilczer W. Introduction to spectroscopy of magnetic resonances. Warszawa: PWN; 2005 (in Polish).

[43] Eaton GR, Eaton SS, Salikhov KM. Foundations of modern EPR. Singapore: World Scientific; 1998.

[44] Pilawa B, Latocha M, Kościelniak M, Pietrzak R, Wachowska H. Oxygen effects in tumor cells during photodynamic therapy. Polish Journal of Environmental Studies. 2006;**15**:160-162.

[45] Latocha M, Pilawa B, Zdybel M, Wilczok T. Effect of laser radiation on free radicals in human cancer G361 cells. Acta Physica Polonica A. 2005;**108(2)**:409-412.

[46] Godechal Q, Ghanen GE, Cook MG, Gallez B. Electron paramagnetic resonance spectrometry and imaging in melanomas: comparison between pigmented and nonpigmented human malignant melanomas. Molecular Imaging. 2013;**12(4)**:218-223. DOI: 10.2310/7290.2012.00037

[47] Godechal Q, Defrense F, Danhier P, Leveque P, Porporato PE, Sonveaux P, Baurain JF, Feron O, Gallez B. Assessment of melanoma extent and melanoma metastases invasion using electron paramagnetic resonance and bioluminescence imaging. Contrast Media & Molecular Imaging. 2011;**6(4)**:282-288. DOI: 10.1002/cmmi.430

[48] Godechal Q, Leveque P, Marot L, Baurain JF, Gallez B. Optimization of electron paramagnetic resonance imaging for visualization of human skin melanoma in various stages of invasion. Experimental Dermatology. 2012;**21(5)**:341-346. DOI: 10.1111/j.1600-0625.2012.01461.x

[49] Vanea E, Charlier N, Dewever J, Dinquizli M, Feron O, Baurain JF, Gallez B. Molecular electron paramagnetic resonance imaging of melanin in melanomas: a proof of concept. NMR in Biomedicine. 2008;**21(3)**:296-300. DOI: 10.1002/nbm.1241

[50] Godechal Q, Gallez B. The contribution of electron paramagnetic resonance of melanoma research. Journal of Skin Cancer. DOI: 10.1155/2011/273280

[51] Romanowska Dixon B, Klonowska A, Plonka P, Sarna T. Identification and characterization of melanin in choroidal melanoma by electron paramagnetic resonance spectroscopy (EPR). Acta Ophtalmologica. 2014;**92(253)**. DOI: 10.1111/j.1755-3768.2014.4247.x

[52] Wolnicka-Glubisz A, Nogal K, Żądło A, Płonka PM. Curcumin does not switch melanin synthesis towards pheomelanin in B16F10 cells. Archives of Dermatology Research. 2015;**307(1)**:89-98. DOI: 10.1007/s00403-014-1523-1

[53] Avendo C, Mendendez JC. Medicinal Chemistry of Anticancer Drugs. 2nd ed. Amsterdam: Elsevier Science; 2015.

[54] Rathod CP, Dhawale SC, Kshirsagar RV. Recent trends in screening and evaluation methods of anticancer drugs. Indo American Journal of Pharmaceutical Research. 2012;**2(5)**:506-515.

[55] Zdybel M, Pilawa B, Buszman E, Wrześniok D. Effect of oxygen on free radicals in DOPA-melanin complexes with netilmicin, diamagnetic Zn(II), and paramagnetic Cu(II). Chemical Physics Letters. 2013;**556**:278-286. DOI: 10.1016/j.cplett.2012.12.004

[56] Wiernek BK, Pilawa B, Zdybel M, Buszman E, Wrześniok D. Interaction of free radicals of DOPA-melanin-streptomycin complexes with paramagnetic oxygen O_2. Journal of Applied Biomedicine. 2014;**12(3)**:161-169. DOI: 10.1016/j.jab.2013.12.002

[57] Beberok A, Zdybel M, Pilawa B, Buszman E, Wrześniok D. EPR characteristics of free radicals in DOPA-melanin-moxifloxacin complexes at ambient level of UVA radiation. Chemical Physics Letters. 2014;**592(1)**:41-46. DOI: 10.1016/j.cplett.2013.12.011

[58] Zdybel M, Chodurek E, Pilawa B. EPR studies of free radicals in A-2058 human melanoma cells treated by valproic acid and 5,7-dimethoxycoumarin. Acta Poloniae Pharmaceutica – Drug Research. 2014;**71(6)**:1066-1072.

[59] Chodurek E, Zdybel M, Pilawa B, Dzierżewicz Z. Examination by EPR spectroscopy of free radicals in melanins isolated from A-375 cells expose on valproic acid and cisplatin. Acta Poloniae Pharmaceutica – Drug Research. 2012;**69(6)**:1334-1341.

[60] Chodurek E, Zdybel M, Pilawa B. Application of EPR spectroscopy to examination of free radicals in melanins from A-375 and G-361 human *melanoma malignum* cells. Journal of Applied Biomedicine. 2013;**11**:173-185. DOI: 10.2478/v10136-012-0023-x

[61] Wolny D, Chodurek E, Dzierżewicz Z. Antiproliferative effect of valproic acid and 5,7-dimethoxycoumarin against A2058 human melanoma cells. Acta Poloniae Pharmaceutica – Drug Research. 2014;**71(6)**:1056-1059.

[62] Chodurek E, Kuśmierz D, Dzierżęga-Lęcznar A, Kurkiewicz S, Stępień K, Dzierżewicz Z. Thermochemolysis as the useful method to assess the purity of melanin isolated from the human *melanoma malignum*. Acta Poloniae Pharmceutica – Drug Research. 2008;**65**:731-734.

[63] Chodurek E, Orchel A, Orchel J, Kurkiewicz S, Gawlik N, Dzierżewicz Z, Stępień K. Evaluation of melanogenesis in A-375 cells in the presence of DMSO and analysis of pyrolytic profile of isolated melanin. The Scientific World Journal. 2012;**2012**:854096. DOI: 10.1100/2012/854096

[64] Ramos P, Pilawa B. EPR examination of free radicals thermally formed in vaselinum flavum. Nukleonika. 2015;**60(3)**:443-447. DOI: 10.1515/nuka-2015-0079

[65] Ramos P, Pieprzyca M, Pilawa B. Effect of microwave power on shape of EPR spectra and application to examination of complex free radical system in thermally sterilized acidum boricum. Acta Poloniae Pharmaceutica - Drug Research. 2016;**73(2)**:291-296.